An Introduction to
the Liquid State

An Introduction to the Liquid State

P. A. EGELSTAFF

Atomic Energy Research Establishment
Harwell, Berkshire, England

ACADEMIC PRESS
London and New York

1967

ACADEMIC PRESS INC. (LONDON) LTD
Berkeley Square House
Berkeley Square
London W.1.

U.S. Edition published by
ACADEMIC PRESS INC.
111 Fifth Avenue
New York, New York 10003

117157.

PRINTED IN GREAT BRITAIN BY
J. W. ARROWSMITH LTD., WINTERSTOKE ROAD, BRISTOL 3

Contents

Acknowledgments

For permission to use, in whole or in part, certain diagrams we are grateful to the following publishers and journals—
American Institute of Physics; W. A. Benjamin, Inc.; Elsevier Publishing Company; International Atomic Energy Agency; *Journal of Chemical Physics; Journal of Mathematics and Physics;* McGraw-Hill Publishing Company; North-Holland Publishing Company; Pergamon Press; *Physica; Proceedings of the Physics Society; Proceedings of the Royal Society; Reports on the Progress of Physics; The Physical Review;* John Wiley and Sons, Inc.

Detailed acknowledgments are given in the captions to illustrations.

Preface

This book has grown out of a lecture course of the same title given in 1965 at the University of Reading and later at AERE. The course was aimed at the postgraduate physicist rather than the specialist in the liquid state, and for this reason it gives a general presentation of the subject at the postgraduate level. At each stage the relationship between the theoretical predictions and the experimental situation is examined, and an effort has been made to focus the discussion on real liquids rather than theoretical models of liquids. In particular, a comparison between liquid metals and rare-gas liquids has been adopted throughout in order to bring out those liquid properties that are independent of the range of the interatomic forces.

The view of liquids taken here is strictly atomistic, each property being related to the details of the atomic motions and the atomic positions. The types of liquids discussed are simple in the sense that the internal motions and structure of molecules are not examined; only the motion of molecules as a whole or the motion of the atoms in a monatomic liquid are considered relevant. No attempt has been made to cover every model or treatment of the liquid state. A simple self-consistent treatment is followed throughout that involves the pair distribution and pair potential functions (i.e., the pair theory of liquids). The basic theme is to show how every property (macroscopic or microscopic) may be calculated from first principles; that is from the forces between atoms calculated ideally via a basic quantum mechanical treatment of electronic behaviour. The predictions of this theory should then be compared to the experimental results in order to test the validity of the approach, but as will be seen this programme is far from complete because the crucial tests are difficult to carry out properly. The aim has been to present the whole subject in a rounded way rather than specialize on one area, and the treatment has been chosen for its physical content rather than mathematical rigour. This method would seem to meet the needs of the experimentalist or of the student needing an introduction. It is hoped that an adequate introduction to the subject has been given which will enable the reader to go on to the more sophisticated theories involving triplet and higher-order correlations or to specialist treatments for particular liquid properties.

The only techniques for measuring atomic positions, motions and forces are radiation or atomic-scattering techniques. Thus considerable emphasis has been given to scattering methods and theory, in-including the basis of the measuring techniques and the significance of the results obtained. In contrast, almost no space has been devoted to the methods of measuring macroscopic properties, although the actual values of the macroscopic constants are discussed at reasonable length. Broadly the book has been divided into two parts, the first dealing with distribution functions involving atomic positions only (Chapters 1–7) and the second part distribution functions involving both position and time (Chapters 8–14). In a general sense the former covers equilibrium properties, and the latter covers transport properties. The whole of these two parts is devoted to classical liquids not too far from the triple point. In order to round out the treatment one Chapter each is devoted to a sketch of the critical region and of quantum liquids.

It has been assumed that the reader has a knowledge of both elementary quantum mechanics and of classical statistical mechanics: for example, it is assumed that he is familiar with the idea of a partition function and with the general ideas of quantum mechanical scattering theory. Much of the material found in a course on statistical mechanics is, because of its general nature, relevant to a course on the liquid state (although the sequence normally followed in statistical mechanics is not appropriate here, and the order taken has been chosen for its relevance to the liquid state). Most of the ground covered is well established and as far as possible references have been given to well known text books. However subject matter that has been compiled from scattered sources is presented from the standpoint of the liquid state, and in order to achieve a uniform presentation some of the more difficult parts have been given in outline only. In particular, the basic quantum mechanical calculation of the pair potential is sketched in outline only.

The style followed is a division of the material into short compact Chapters, and the sub-division of each Chapter into short Sections, each of which covers one idea. It is hoped that this will provide a student with numerous "signposts" as the subject unfolds. In addition "progress reports" are given at the beginning of a number of Chapters in order to indicate the problems covered or those remaining. Finally the mathematical statement has been preferred to the verbal statement where this may be done. The reasons are twofold, first for the sake of definiteness and secondly to provide a good key to the references in which fuller and deeper treatments may be found. A list of symbols followed generally throughout the book is given after this preface. In addition, short lists are given at the end of each Chapter to indicate those symbols which are special to that Chapter.

I would like to thank Bill Mitchell of Reading for encouraging me to undertake this course and Otto Eder, whose notes were of help in writing the present manuscript. I am indebted to Peter Schofield and Philip Hutchinson of AERE for many helpful remarks on the material in this book and to John Enderby of Sheffield for advice on Chapter 4. I am particularly grateful to Philip Hutchinson for carefully reading and checking the manuscript and proofs. Finally I would like to thank the Academic Press for their co-operation in the rapid production of this book.

Abingdon, Berks.
May, 1967   PETER EGELSTAFF

General List of Symbols

A	Ratio of nuclear to neutron masses	$h(r)$	"Total" correlation function, i.e. $\{g(r)-1\}$
$A(Q, \omega)$	Fourier transform of Imag $G(r,\tau)$ —equation (14.15a)	$\hbar$	Planck's constant divided by 2π
B	Instantaneous bulk modulus	i, j, l	Indices denoting ith(jth or lth) particle
b	Bound atom scattering length (neutrons)	$I(Q, \tau)$	Intermediate scattering function, or Fourier transform of $G(r, \tau)$ with respect to $\mathbf{r}$
b_{coh}	Bound atom coherent scattering length (neutrons)	$\mathscr{I}$	Imaginary part of
b_e	Scattering length for electrons	k	Boltzmann's constant
b_{incoh}	Bound atom incoherent scattering length (neutrons)	k	Wave vector of scattered radiation
b_L	Scattering length for light	k_f	Fermi radius in a metal
b_X	Scattering length for X-rays	k_0	Wave vector of incident radiation
c	Velocity of light		
c	Velocity of sound	l	Angular momentum quantum number
$c(Q)$	Fourier transform of $c(r)$		
$c(r)$	Direct correlation function	l	Length of diffusive step
C_V	Specific heat at constant volume	$\mathscr{L}$	Grand Partition function
C_P	Specific heat at constant pressure	M	Molecular mass
		M^*	Effective mass for diffusion
$d\{N\}$	$d\mathbf{r}_1, d\mathbf{r}_2, d\mathbf{r}_3 \ldots d\mathbf{r}_N$	m	Reduced mass
D	Diffusion constant	m_e	Electron mass
D_T	Thermal diffusivity $= \lambda/C_V\rho$	m_n	Neutron mass
$D_{T'}$	Thermal diffusivity $= \lambda/C_P\rho$	N	Number of atoms or molecules (usually in volume V)
E	Energy		
e	Charge on electron	$n^{(1)}, n^{(2)}$	
F	Free energy	$\ldots n^{(r)}$	Molecular distribution functions for r molecules
$F_x(Q)$	X-ray form factor—equation (6.20)	P	Pressure
f_{ij}	$\exp\{-u(r_{ij})/kT\}-1$	p	Momentum of an atom
G	Rigidity modulus	Q	Wave vector difference, $|\mathbf{k}_0-\mathbf{k}|$
$g(r)$	Pair distribution function	Q_N	Partition function for N molecules
$G(r, \tau)$	Total space–time correlation function	Q_0	Value of Q at major peak of $S(Q)$
$G_s(r, \tau)$	Self part of space–time correlation function	q	Wave number of co-operative mode
$G_x(r)$	Correlation function for electron density	q	Momentum of atom, or collective mode divided by $\hbar$
$G_e(r)$	Correlation function for charge density	$\mathbf{r}, \mathbf{r}_1$, etc.	Position vectors
H	Hamiltonian	r_c	Bohr radius of electron
$h(Q)$	Fourier transform of $h(r)$	r_m	Range in Buckingham potential

$\mathscr{R}$	Real part of	$\tilde{z}(\omega)$	Spectral density of a correlation
S	Entropy		function (with subscript, spec-
$\mathbf{s}$	Relative position vector		tral density of stress correlation)
$S(Q)$	Liquid structure factor, equal to	β	Amplitude of repulsive term in
	$\int S(Q, \omega)d\omega$		Harrison potential
$S(Q, \omega)$	Scattering law	γ'	Time decay constant for sto-
T	Temperature (°K)		chastic force
$\mathbf{t}$	Relative position vector	γ	Friction constant
$U\{N\}$	Potential energy for N atoms at	ε	Well depth in $u(r)$
	positions $r_1, r_2, r_3 \ldots r_N$	$\varepsilon(q)$	Dielectric function
$u(r)$	Pair potential developed be-	ζ	Bulk viscosity
	tween any pair of molecules	η	Shear viscosity
V	Volume of liquid	η	Density parameter for hard-
$V(Q, \omega)$	General viscosity coefficient		sphere fluid
$V(r)$	Scattering potential (usually	θ	Scattering angle
	nuclear)	λ	Thermal conductivity
$\mathbf{v}$	Particle velocity	μ	Chemical potential
$w(r)$	Pseudo-potential for a metallic	ρ	Number density of liquid (N/V)
	ion	$\boldsymbol{\sigma}$	Stress tensor
$w^0(r)$	Unscreened pseudo-potential	σ	Range in L.J. potential, also
	for a metallic ion		hard-sphere radius
$w(\tau)$	Mean square displacement in	σ_T	Total cross-section
	$G_s(\mathbf{r}, \tau)$	τ	Time difference in Van Hove
Z_N	Configuration integral for N		correlation function
	molecules	τ_0	Lifetime of diffusive step
Z	Valence of an atom, or number	$\phi(r)$	Potential of mean force
	of conduction electrons per	χ_S	Adiabatic compressibility
	atom in a metal	χ_T	Isothermal compressibility
$z(\tau)$	Velocity correlation function	$\psi(r)$	Wave function
	(with subscript, stress correla-	Ω	Angular variable
	tion function)		

CHAPTER 1

General Properties of Liquids

1.1. Introduction

At low pressures, matter usually exists either as a dense solid or as a dilute vapour. For each of these states there is an ideal model which is a good approximation to reality and forms the basis for theoretical discussion. These are the ideal crystal lattice and the ideal gas models, respectively: in the former, emphasis is on structural order modified slightly by the thermal motion of the atoms while the latter model describes the thermal motion of the atoms on the basis of random atomic positions and motions. At higher pressures a third state of matter appears—the liquid state. This state occurs over a temperature range that separates the regions occupied by the solid and vapour states. At present there is no ideal model which gives a good approximation to the liquid state. If such a model is found it would (in contrast to the crystal and gas models) need to cover structural and thermal properties with equal emphasis.

To specify a liquid it is important to refer to a diagram of state that describes the relationships between pressure, volume and temperature. A diagram of this kind for a typical monatomic liquid (e.g. argon) is given in Fig. 1.1, part (a) being a P–T diagram while part (b) is a P–V diagram. The shaded regions on these diagrams indicate the area being discussed here (apart from the last two Chapters). In this region the volume change on melting is small compared with the volume change on evaporation. As will be shown later, this implies that the latent heat of fusion is small compared to the latent heat of vaporization, so that the interatomic binding in the shaded region of Fig. 1.1 is similar to that in a solid. The temperatures at which most elements are liquid are high enough that the system may be discussed from a classical point of view. This view will be adopted for the present and examined in detail in Chapter 9.

It is usual to classify liquids according to the types of interatomic forces present (the different forces are discussed in Chapters 3 and 4). Six different kinds of liquids may be identified—

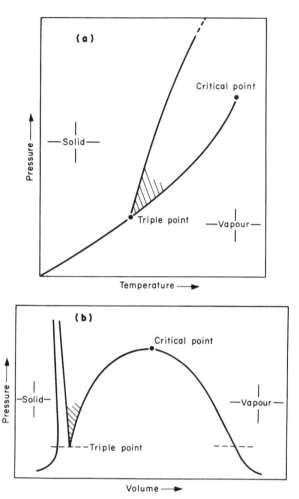

FIG. 1.1(a). *P–T* diagram for a typical element: the shaded region indicates the part of the diagram considered here. (b). *P–V* diagram for a typical element: the shaded region indicates the part of the diagram considered here. In both diagrams the full lines indicate the boundaries of the given states of matter and limits (triple and critical points) of the liquid state are marked. (These diagrams are not to scale.)

(a) Spherical molecules (e.g. A, CH_4) interacting with van der Waals forces.

(b) Homonuclear diatomic molecules (e.g. H_2, N_2)—similar to (a) but the effect of electrical quadrupole moments must be included.

(c) Metals (e.g. Na, Hg)—long-range coulomb forces.

The discussion given here will apply mainly to liquids in these three classes, and particularly to the monatomic ones (e.g. A, Na).

(d) Polar liquids (e.g. HBr)—molecules with electric dipole moments.

(e) Associated liquids or hydrogen bonded liquids (e.g. water, glycerol).

(f) Liquids composed of large molecules or compounds that have important internal modes of motion.

The latter three classes involve phenomena that are not necessarily connected with the liquid state, and hence their treatment is more complex than for the former three. They will not be considered in detail, but nevertheless much of the discussion given here will apply to them.

1.2. Comparison of Liquid Argon and Liquid Sodium

It is of particular interest to compare type (a) with type (c) liquids since the long-range forces are much weaker in (a) compared to (c). Using argon and sodium as representative examples, a comparison of several properties is given in Table 1.1. The first three lines of the Table show that the liquid range is much wider for the metals than the rare gases. This is a direct consequence of the difference in attractive forces. Another feature of interest is the ratio of the maximum attractive potential (or the well depth discussed in Chapters 3 and 4) compared to the temperature of the triple point (Fig. 1.1). This ratio is shown in row 4 and is of order unity in both cases, which shows that it is possible for an atom to escape from a potential well formed by its neighbours and move somewhat freely through any unoccupied volume. The lower triple points for the rare gases are thus associated with shallower potential wells. In the discussion of Fig. 1.1 it was pointed out that there is only a small volume change on melting, and this is shown by the densities listed in line 5. In some unusual cases (e.g. Bi, Ga, H_2O) the liquid density is greater than the solid, but this is thought to be due to an unusually open crystal structure occurring in the solid state rather than a peculiarity in the liquid state (at least for points reasonably far from the triple point). The latent heats shown in line 6 confirm that the binding energy in the liquid is similar to that in the solid.

Heat capacities are shown in line 7, and the theoretical values for the ideal gas and the harmonic oscillator are given for comparison. It can be seen that solid argon and both solid and liquid sodium have values similar to those for the harmonic oscillator. On the other hand, liquid argon does not have a heat capacity close to either

TABLE 1.1

Comparison between physical properties in the liquid and solid state of two simple materials

Property	Argon		Sodium		Comments
	Solid[1]	Liquid[2]	Solid[3]	Liquid[4]	
1. Melting and boiling point at 1 atm.	84°K (m.p.)	87·5°K(b.p.)	371°K(m.p.)	1150°K(b.p.)	
2. Liquid range[5] at 1 atm.	—	3°K	—	782°K }	Note: greater range for metal
3. Liquid Range (C.P./T.P.)	—	1·8	—	7·5 }	
4. Ratio					
Well depth/triple point		≃1·5		≃1·2	Similar result in both cases
5. Density (g cm⁻³)	1·636	1·407	0·951	0·927	Larger change on melting for rare gas
6. Latent heat of vaporization (cal g mol⁻¹)	1850(0°K)	1357·5(87·5°K)	24,000	23,000	Latent heat for vaporization similar in solid and liquid
7. Heat capacity per atom					Perfect atomic gas
C_v(ergs/°K)	2·89k	2·32k	3·1k	3·4k	$C_v = 1·5k$
C_p(ergs/°K)	3·89k	5·06k(87°K)	3·3k	3·8k	$C_p - C_v = k$
(k = Boltzmanns constant)					Harmonic oscillator $C_v = 3k;\ C_p = C_v$
8. Compressibility (cm² dyne⁻¹)					Taken on either side of M.P.
Isothermal	0·95 × 10⁻¹⁰	2·0 × 10⁻¹⁰	1·7 × 10⁻¹¹	1·9 × 10⁻¹¹	Note differences between metal and rare gas
Adiabatic	0·55 × 10⁻¹⁰	0·92 × 10⁻¹⁰	1·6 × 10⁻¹¹	1·8 × 10⁻¹¹	
9. Velocity of sound (m s⁻¹)	1300	874	2600	2500	Similarity of sound velocity for metal in solid and liquid states
10. Viscosity (poise)		2·8 × 10⁻³ (86°K)		6·8 × 10⁻³	Diffusion and Viscosity coefficients similar for all liquids
11. Diffusion coefficient (cm² sec⁻¹)	10⁻⁹	1·6 × 10⁻⁵ (86°K)	1·9 × 10⁻⁷ (371·5°C)	4·3 × 10⁻⁵	
12. Heat conductivity cal sec⁻¹ cm⁻¹ °K⁻¹	7·1 × 10⁻⁴ (84°K)	2·9 × 10⁻⁴ (87°K)	0·322	0·203	High conductivity in metals due to electron transport

[1] at 80°K if not stated differently
[2] at 84·5°K if not stated differently
[3] at 371°K if not stated differently
[4] at 373°K if not stated differently
[5] Distance between solid and liquid line in P–T diagram in °K at 1 Atm.

model. The larger value of C_P is of particular interest and will be commented upon later.

Compressibilities are given in line 8. It is notable that the absolute value of the compressibility and its fractional change on melting are both greater for argon than sodium. This difference is characteristic and indicates that there is a stronger repulsion in a metal than a rare gas (due to ion–ion interactions rather than atom–atom interactions). The behaviour of the velocity of sound (line 9) follows from the behaviour of the compressibility.

Values of the viscosity and diffusion coefficients are given in lines 10 and 11. Both coefficients change by several orders of magnitude on melting,* but both liquids have similar values near the triple point. The values given in the Table seem to be characteristic of the liquid state, and provide the best empirical definition of a liquid (i.e. a medium exhibiting a free surface and having diffusion and viscosity coefficients in this range). A major difference between metals and non-metals is illustrated by the values of the thermal conductivity in line 12. In a metal the conduction electrons produce a conductivity several orders of magnitude larger than that in a non-metal. However in both cases the conductivity falls on melting due to the greater disorder in the liquid state.

The usual difficulty in defining a liquid arises because the application of words like elastic, compressible or viscous to a medium depends upon the rate at which forces are applied. For example, the "bouncing silicones" may be deformed easily in the hand but when dropped bounce from the floor like a rubber ball. Such materials are made of complex silicone molecules, although simple liquids would behave in the same way if subjected to sufficiently high frequency (alternating) forces. This example indicates the similarity between liquids in class (f) and the other classes; however the large viscosity usually associated with this class would exclude them from the definition mentioned in the previous paragraph and as stated above, the behaviour of these liquids is beyond the scope of this book. The definition proposed here is convenient because it limits the discussion to the simpler classes of liquid.

1.3. Some Thermodynamic Relationships

Liquids do not fall into any simple classification, and often must be treated as general "many body" systems. For this reason it is

---

* Solids show a gradual yielding to shear forces (creep) and their effective viscosity may be calculated from this effect.

frequently necessary to base the discussion upon the general laws of statistical thermodynamics. Two examples will be given that amplify the discussion of lines 5 to 9 of Table 1.1.

As a starting point consider the equation for the total differential of the energy (E) of the system (Landau and Lifshitz,[1] equation 24.5)—

$$dE = T\,dS - P\,dV + \mu\,dN \tag{1.1}$$

where T, S, P, V and N are the temperature, entropy, pressure, volume and number of particles in the system. The quantity μ is the chemical potential and is defined by equation (1.1) or—

$$\mu = \left(\frac{\partial E}{\partial N}\right)_{S,V} \tag{1.2}$$

In addition it may be shown that (Landau and Lifshitz,[1] equation 24.12)—

$$N\,d\mu = -S\,dT + V\,dP \tag{1.3}$$

and also that if there are two states (1 and 2) in equilibrium the chemical potentials are equal (Landau and Lifshitz,[1] equation 25.1), or—

$$\mu_1(P, T) = \mu_2(P, T) \tag{1.4a}$$

Differentiation of this equation with respect to T yields (remembering that P is not an independent variable but is a function of temperature determined by equation 1.4a)—

$$\left(\frac{\partial \mu_1}{\partial T}\right)_P + \left(\frac{\partial \mu_1}{\partial P}\right)_T \frac{dP}{dT} = \left(\frac{\partial \mu_2}{\partial T}\right)_P + \left(\frac{\partial \mu_2}{\partial P}\right)_T \frac{dP}{dT}$$

By virtue of equation (1.3) this equation may be rewritten—

$$S_2 - S_1 = (V_2 - V_1)\frac{dP}{dT} \tag{1.4b}$$

and the latent heat (q) of the transition is—

$$q = (S_2 - S_1)T = (V_2 - V_1)T\frac{dP}{dT} \tag{1.5}$$

(this equation is known as the Clapeyron–Clausius equation). Thus the latent heat is proportional to the volume difference (per atom) of the two phases, and the similarity of the results in lines 5 and 6 of Table 1.1 can be understood. Also, since the volume difference between liquid and gas vanishes at the critical point (Fig. 1.1), the latent heat vanishes to the same order. Thus in the critical region, the latent heat for vaporization is much less than that for fusion (at constant pressure) and hence the liquid binding energy is smaller than that of the solid at the same pressure.

Now consider a large volume of liquid and compare its properties (temperature, entropy, etc.) with those of a much smaller volume. The values appropriate to the small volume fluctuate about the values associated with the large volume. The magnitude of the thermodynamic fluctuations may be calculated[1] from the probability (w) of a reversible modification. If R is the minimum work required to produce this effect then (Landau and Lifshitz,[1] equation 111.1)—

$$w \propto \exp -(R/kT) \tag{1.6}$$

where k is Boltzmann's constant. But R may be shown (Landau and Lifshitz,[1] equation 20.8) to be (cf. equation 1.1)—

$$R = \Delta E - T\Delta S + P\Delta V \tag{1.7}$$

where Δ indicates a fluctuation. Then assume that the sizes of the fluctuations are small and that it is possible to expand ΔE as a Taylor series in ΔS and ΔV, i.e.—

$$\Delta E = \frac{\partial E}{\partial S}\Delta S + \frac{\partial E}{\partial V}\Delta V + \frac{1}{2}\left[\frac{\partial^2 E}{\partial S^2}(\Delta S)^2 + 2\frac{\partial^2 E}{\partial S \partial V}\Delta S \Delta V + \frac{\partial^2 E}{\partial V^2}(\Delta V)^2\right] + \text{etc.}$$

Finally by remembering that—

$$T = \left(\frac{\partial E}{\partial S}\right)_V \quad \text{and} \quad P = \left(\frac{\partial E}{\partial V}\right)_S$$

the expression for R may be reduced to—

$$R = \tfrac{1}{2}(\Delta S \cdot \Delta T - \Delta P \cdot \Delta V) \tag{1.8}$$

From this expression the separate fluctuations in the four quantities may be calculated. First take S and P as independent variables, and express ΔT and ΔV in terms of them (again assuming that the fluctuations are small)—

$$\Delta V = \left(\frac{\partial V}{\partial P}\right)_S \Delta P + \left(\frac{\partial V}{\partial S}\right)_P \Delta S$$

$$= -V\chi_S \Delta P + \left(\frac{\partial T}{\partial P}\right)_S \Delta S \tag{1.9}$$

since[1] $(\partial V/\partial S)_P = (\partial T/\partial P)_S$ and the adiabatic compressibility (χ_S) is defined as—

$$\chi_S = -\frac{1}{V}\left(\frac{\partial V}{\partial P}\right)_S$$

Also—

$$\Delta T = \left(\frac{\partial T}{\partial S}\right)_P \Delta S + \left(\frac{\partial T}{\partial P}\right)_S \Delta P = \frac{T}{C_P}\Delta S + \left(\frac{\partial T}{\partial P}\right)_S \Delta P \qquad (1.10)$$

since the specific heat at constant pressure C_P is defined by—

$$C_P = T\left(\frac{\partial S}{\partial T}\right)_P$$

Substituting equations (1.9) and (1.10) into equation (1.8) gives—

$$2R = (\Delta S)^2 \frac{T}{C_P} + (\Delta P)^2 V\chi_S \qquad (1.11)$$

Now by using equation (1.11) in equation (1.6), the probability w is recognized to be a Gaussian function of two statistically independent variables (Yule and Kendell[2]). The statistical independence implies that—

$$\langle \Delta S . \Delta P \rangle = 0 \qquad (1.12a)$$

and from (1.11) the mean-square deviations are—

$$\left.\begin{array}{l} \langle (\Delta S)^2 \rangle = kC_P \\ \langle (\Delta P)^2 \rangle = kT/V\chi_S \end{array}\right\} \qquad (1.12b)$$

where the $\langle --- \rangle$ denotes an average value.

An identical calculation may be carried out taking T and V as independent variables, and then equation (1.8) becomes—

$$2R = (\Delta T)^2 \frac{C_V}{T} + (\Delta V)^2 V\chi_T \qquad (1.13)$$

where C_V is the specific heat at constant volume and χ_T is the isothermal compressibility. This equation again gives R as a Gaussian function of two statistically independent variables,[2] so that the following results are obtained—

$$\left.\begin{array}{l} \langle \Delta T . \Delta V \rangle = 0 \\ \langle (\Delta T)^2 \rangle = kT^2/C_V \\ \langle (\Delta V)^2 \rangle = kT/V\chi_T \end{array}\right\} \qquad (1.14)$$

These results may now be used to discuss lines 7–9 of Table 1.1. The relatively large change of C_P for argon on melting is now seen to be associated (equation 1.12) with an increase in the entropy fluctuations, and the decrease of C_V arises (equation 1.14) from an increase in the temperature fluctuations. Both of these effects are larger f_ argon than sodium. In contrast, the volume and pressure fluctuations decrease on melting, since the compressibilities (of line 8) increase. Again these effects are larger for argon than sodium.

The increase in entropy and temperature fluctuations is suggestive of an increase in disorder and thermal motion in the liquid compared to the crystalline solid. However, the decrease in volume and pressure fluctuations suggests that the liquid is macroscopically "smoother" than the crystalline solid. It is hard to avoid the conclusion that these two effects are interconnected.

1.4. Theories of the Liquid State

As remarked in Section 1.1, there is no theory of the liquid state in the sense (i.e. based on a simple model) that theories of the solid and gaseous states are known. However in another sense there is a well defined theory of the liquid state. It will be shown how almost all the properties of a liquid may be calculated from the hypothesis that the interatomic forces and energies in a classical liquid are dominated by the sum of pair interactions. That is if three or more atoms are close together in a dense liquid, the energy of the system is the sum of the energies obtained by taking the atoms in pairs. Thus "3-body" or higher-order forces are neglected in this theory. From this "pair theory" of the liquid state, expressions for the equation of state, for example, are readily obtained.

A great deal of work has been published on calculations of the quantities that appear in these equations and from which numerical (P, V, T) values are obtained to compare with experiment. Sometimes in the literature, the phrase "theory of liquids" is used solely to describe a particular version of such calculations, and in this sense there are many theories of liquids. In this book the phrase "a theory of liquids" is interpreted in a wider sense than just meaning a model for (P, V, T) calculations. Thus the "pair theory of liquids" means the attempt to derive expressions for all liquid properties (equilibrium, transport, macroscopic and microscopic) on the basis of the pair approximation. It will be shown that there are substantial difficulties in using these expressions to obtain accurate numerical predictions of almost all properties, so that the pair theory has not yet been adequately tested. Moreover a full test of the pair theory should indicate where it is able to predict experimental data adequately and where it fails to do so. In some cases (the most important of which are the diffusion and viscosity coefficients) the pair theory has not been developed to the point where rigorous expressions may be derived. In these cases approximate expressions (based on a model which is used in addition to the pair theory) are commonly employed.

Thus the models are used in two ways (a) to obtain mathematical expressions for physical quantities and (b) to obtain numerical estimates for the functions appearing in these expressions. It is important to distinguish between these two uses of models. It is also important to distinguish between cases where a single hypothesis (e.g. the pair approximation) has been used and cases where a model is required in addition to the hypothesis. The layout of this book has been made with these requirements in mind. Thus the body of the material is concerned with the problems of (a) deriving exact pair-theory expressions and of (b) measuring directly the quantities which appear in these expressions. Sections in which a model is employed are indicated and some comments made on the sphere of application of the model.

Thus the object of this book is to present the pair theory of liquids, covering the principal physical properties of liquids and showing how the basic microscopic functions may be measured and the theory tested in principle.

1.5. Methods of Studying the Liquid State

In the main there are four types of information which it is necessary to study. These are (i) the equilibrium properties, (ii) the transport properties, (iii) the macroscopic data and (iv) the microscopic data. In class (i) the positions and energies of the atoms are important, whereas in class (ii) the motion of the atoms is the significant feature. Clearly both of these classes are extremely important for the study of the liquid state. For each case it is possible to make a sub-division into (iii) macroscopic and (iv) microscopic data. It will be shown that the macroscopic data involve integrations over the microscopic data and consequently are much less sensitive to the details of a liquid-state theory. To offset this point it should be emphasized that the macroscopic data can be measured usually to a higher accuracy ($\sim 1\%$) than the microscopic data ($\sim 10\%$), and in this way it can in principle form a good test of the theory. However with the present state of knowledge this higher accuracy is not required, since it is necessary to obtain the correct general form and magnitude of (e.g.) the equation of state, before examining the detailed behaviour. Consequently it is useful to concentrate on the microscopic data as the most powerful method of studying the liquid state at present.

Microscopic studies of the liquid state involve scattering processes of a number of types. Thus in the following Chapters, special attention is devoted to the atomic interactions and the various scattering techniques for studying them.

Symbols for Chapter 1

q Latent heat at phase change
R Minimum work required for a thermal fluctuation
w Probability of a thermal fluctuation occurring

Molecular Distribution Functions and the Equation of State

2.1. Molecular Distribution Functions

The significance of the diagrams of state (Fig. 1.1a, b) has been stressed in Chapter 1. In Chapter 2 the basic equations used to discuss these diagrams are derived in outline (for a rigorous derivation the reader should consult Huang[3] and Hill[4]). Two general forms of the equation of state will be obtained, and some of the expansions of these equations will be considered. In addition, the expressions for the internal energy and the specific heat at constant volume will be derived.

The starting point of these calculations is the molecular pair distribution function, which is a particular member of a general class of distribution functions. These functions describe the probability of occurrence of a particular arrangement of molecules (in thermal equilibrium) in the liquid. In one method, a system containing a fixed number (N) of particles in a fixed volume (V) is considered and the term "canonical ensemble" is used to denote it. The calculations are carried through in the limit $N \to \infty$ and $V \to \infty$ while the ratio $N/V = \rho$ remains a constant, (in addition the surface-to-volume ratio should approach zero as the volume becomes infinite). In order to distinguish a calculation of this kind a subscript, N will be used on the distribution functions. Since the number N is fixed, equation (1.1) can be used to show that the chemical potential does not appear in such calculations.

The above method does not cover those cases in which the fluctuation in the number of particles (N) has to be studied. In such problems a "grand canonical ensemble" is used where the liquid is divided into a number of sub-systems of volume V which are in thermal equilibrium and the number of atoms in each sub-system is a thermodynamic variable. Equation (1.1) shows that a calculation of this type involves the chemical potential, and thus is easily distinguished from calculations involving the canonical ensemble. However, a grand canonical ensemble calculation becomes equivalent to the canonical ensemble calculation in the infinite volume limit only (where $\langle N \rangle / V \to \rho$).

Before discussing the pair distribution function itself, the general definitions of molecular distribution functions will be given. In the canonical ensemble the simplest distributions are—

$n_N^{(1)}(\mathbf{r}_1) \, d\mathbf{r}_1$: is proportional to the probability of finding a molecule at $\mathbf{r}_1$ in a volume element $d\mathbf{r}_1$.

$n_N^{(2)}(\mathbf{r}_1, \mathbf{r}_2) \, d\mathbf{r}_1 \, d\mathbf{r}_2$: is proportional to the probability of finding a molecule at $\mathbf{r}_2$ in a volume element $d\mathbf{r}_2$ if, at the same time, there is a molecule at $\mathbf{r}_1$ in volume element $d\mathbf{r}_1$ (irrespective of where the remaining molecules are).

Clearly this scheme can be extend to any order $n_N^{(n)}(\mathbf{r}_1 \ldots \mathbf{r}_n)$. If it is assumed that the potential energy of the system depends only upon the co-ordinates of the N molecules (i.e. the effects of molecular shape are neglected, or are zero as for the liquids (a) and (c) of Section 1.1), then the Hamiltonian can be written—

$$H = \sum_i^N \frac{1}{2M} p_i^2 + U\{N\} \tag{2.1}$$

where M is the molecular mass, p_i is the momentum of the ith molecule, U is the potential energy and $\{N\}$ denotes $(\mathbf{r}_1 \ldots \mathbf{r}_N)$. The probability $n_N^{(n)}$ is then defined (Rice and Gray,[5] equation 2.4.1) in terms of the potential energy $U\{N\}$ as—

$$n_N^{(n)}(\mathbf{r}_1 \ldots \mathbf{r}_n) = \frac{N!}{(N-n)!} \frac{\int_V \ldots \int \exp -\dfrac{U\{N\}}{kT} \, d\{N-n\}}{Z_N} \tag{2.2}$$

where—

$$Z_N = \int_V \ldots \int \exp -\frac{U\{N\}}{kT} \, d\{N\} \tag{2.3}$$

Equation (2.2) is seen to consist of two factors. First there is a statistical factor, $N!/(N-n)!$, which is the number of ways of placing n molecules (taken from N) in the volume element $d\{n\}$. Secondly there is the sum of the probabilities of finding the remaining $(N-n)$ molecules in any configuration [which is determined by the total potential energy $U\{N\}$]. Calculations in the canonical ensemble are carried through with a finite N in the definition (2.2) and then the limit $N \to \infty$ is taken in the final expression (equation 2.18 is an example). There are some simple relationships between these functions which follow from equations (2.2) and (2.3). For example, $n_N^{(1)} = N/V$ for an isotropic system like a liquid, and—

$$\int_V n_N^{(2)}(\mathbf{r}_1, \mathbf{r}_2)\, d\mathbf{r}_2 = n_N^{(1)}(\mathbf{r}_1) \cdot (N-1) = \frac{N(N-1)}{V} \tag{2.4}$$

is proportional to the probability of finding a molecule in $d\mathbf{r}_1$ at $\mathbf{r}_1$ and the other $(N-1)$ molecules somewhere in the volume V. This integral relationship is easily generalized to $n_N^{(n)}$.

In the grand canonical ensemble a collection of sub-systems covering all possible numbers (N) of molecules is considered. The probability of finding any particular N is proportional to $e^{\mu N/kT}$, (μ is the chemical potential). For convenience this term is often written z^N where—

$$z = e^{\mu/kT} \tag{2.5}$$

is called the activity or "fugacity". Thus in the grand canonical ensemble the definition (2.2) is broadened (Rice and Gray,[5] equation 2.4.15)—

$$n^{(n)}(\mathbf{r}_1 \ldots \mathbf{r}_N) = \frac{\sum\limits_{N \geq n}^{\infty} \dfrac{z^N}{(N-n)!\Lambda^{3N}} \int_V \ldots \int \exp -\dfrac{U\{N\}}{kT}\, d\{N-n\}}{\mathscr{L}} \tag{2.6a}$$

where Λ^3 is a volume element discussed in Section 2.3, and—

$$\mathscr{L} = \sum\limits_{N \geq 0}^{\infty} \frac{z^N Z_N}{N!\Lambda^{3N}} \tag{2.6b}$$

Note (by comparing equations 2.2 and 2.6) how the statistical weight $N!(N-n)!$ has been split into its component factors. From this expression it is easy to show that—

$$\int n^{(n)}\{n\}\, d\{n\} = \frac{1}{\mathscr{L}} \sum\limits_{N \geq n} \frac{N!}{(N-n)!} \cdot \frac{z^N Z_N}{N!} \tag{2.7a}$$

$$= \left\langle \frac{N!}{(N-n)!} \right\rangle_{\text{over all } N} \tag{2.7b}$$

which is the analogue of equation (2.4). This result follows from the fact that $z^N Z_N/N!$ is proportional to the probability of finding N molecules in a particular member of the grand canonical ensemble. $z^N Z_N/N!$ is simply the product of (i) the chance z^N of placing N particles in the sub-system with (ii) the chance Z_N of all possible spatial configurations of N particles (as given by the potential energy) and finally (iii) divided by the number of ways of choosing N molecules (as the molecules are individually indistinguishable).

2.2. The Pair Distribution Function

In this Section some properties of the two molecule distribution function, $n^{(2)}$, are discussed. A particularly useful result follows from (2.7) for the case of $n = 2$, i.e.—

$$\int_V n^{(2)}(\mathbf{r}_1, \mathbf{r}_2)\, d\mathbf{r}_1\, d\mathbf{r}_2 = \langle N^2 \rangle - \langle N \rangle \qquad (2.8)$$

This result will be used extensively in later Sections. Also for a stationary liquid in thermal equilibrium $n^{(2)}$ cannot depend upon the choice of $\mathbf{r}_1$ (neglecting surface effects) and therefore it depends upon the difference $\mathbf{r} = \mathbf{r}_1 - \mathbf{r}_2$ alone. Further, since the liquid is a macroscopically isotropic body, the direction of $\mathbf{r}$ is unimportant. Thus $n^{(2)}$ depends only upon the scalar $|\mathbf{r}|$.

Now the two particle distribution function is often written as a "pair distribution function" $g(r)$ normalized to unity at large r, and defined by—

$$g(r) = \left(\frac{V}{N}\right)^2 n^{(2)}(r) \qquad (2.9a)$$

Thus $g(r)$ is the probability of finding a molecule at r if there is a molecule at the origin, and it is normalized to unit probability at large r.

It is sometimes useful to define a "potential of mean force" $\phi(r)$ by the relationship—

$$g(r) = \exp\left\{-\frac{\phi(r)}{kT}\right\} \qquad (2.10a)$$

From the definition (2.6), it can be shown that if $\phi(r)$ is differentiated with respect to the coordinates of molecule 1, then (Rice and Gray,[5] equation 2.6.3)—

$$\frac{\partial \phi(r)}{\partial \mathbf{r}_1} = -\frac{kT\, \partial g(r)/\partial \mathbf{r}_1}{g(r)}$$

$$= \frac{\displaystyle\sum_{N \geq 2} \frac{z^N}{(N-2)!\, \Lambda^{3N}} \int_V \cdots \int \exp\left(-\frac{U\{N\}}{kT}\right)\frac{\partial U\{N\}}{\partial \mathbf{r}_1}\, d\{N-2\}}{\displaystyle\sum_{N \geq 2} \frac{z^N}{(N-2)!\, \Lambda^{3N}} \int_V \cdots \int \exp\left(-\frac{U\{N\}}{kT}\right) d\{N-2\}}$$

$$(2.10b)$$

Thus $\phi(r)$ is rightly termed the potential of mean force, as equation (2.10b) gives the normalized average force on molecule 1 if molecule 2 is held fixed at $\mathbf{r}$. In the limit of a dilute gas, it is clear that $\phi(r) = u(r)$ [$u(r)$ is defined at equation 2.17].

Another definition of $g(r)$ is sometimes used in the literature, namely—

$$\rho g(r) = \frac{1}{N} \left\langle \sum_{i \neq j} \delta(\mathbf{r} + \mathbf{r}_i - \mathbf{r}_j) \right\rangle \tag{2.9b}$$

where $\mathbf{r}_i$, $\mathbf{r}_j$ are the positions of the ith and jth molecules, respectively. This definition is discussed in Section 8.1, but will be used in Section 6.3. The equivalence of this definition and the definition at equation (2.9) follows from the normalization of the δ function.[6,7]

The general shape of $g(r)$ for a monatomic liquid is shown at Fig. 2.1. At low r it is almost zero due to the high energies required to

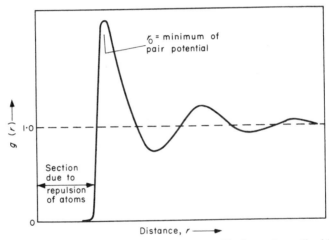

FIG. 2.1. Typical form for the pair distribution $g(r)$; for a dense liquid (Fig. 1.1) the principal peak of $g(r)$ occurs at a position close to the principal minimum of $u(r)$. Typical forms for $u(r)$ are shown in Fig. 3.1.

force atoms to overlap. At a distance roughly equal to the atomic diameter, there is a pronounced peak in $g(r)$, which denotes a sphere of nearest neighbours. At higher values of r there are oscillations representing more distant neighbours. These oscillations decrease in amplitude with increasing r, and eventually $g(r)$ approaches the (unit) mean density of the system. Bernal[8] has pointed out that for simple liquids the general shape of $g(r)$ can be understood from geometrical packing conditions. Figure 2.2 shows how the liquid is built up according to this argument: the atoms are randomly stacked together, except that a minimum distance (r_1) between atoms is imposed. From a model of this type Bernal has computed a $g(r)$ which is very similar to that shown at Fig. 2.1.

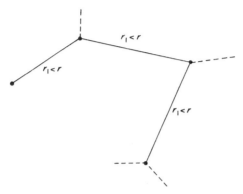

FIG. 2.2. Illustration of how $g(r)$ can be built up by random packing (Bernal). Each full circle represents the position of an atom; the positions are chosen at random except that the distances between atoms must be greater than a minimum distance r_1. The dashed lines indicate possible directions for other atoms.

2.3. The Internal Energy (E) of a Liquid

As an illustration of the usefulness of the molecular distribution function, $n_N^{(2)}$, two basic properties of a liquid will be calculated in the canonical ensemble. The first property is the internal energy, and the starting point is the "partition function" for a canonical ensemble. There is a well known relationship (Landau and Lifshitz,[1] equations 15.5 and 31.3) between the internal energy and this partition function (Q_N), namely—

$$E = kT^2 \left(\frac{\partial \log Q_N}{\partial T} \right)_{N,V} \tag{2.11}$$

so that if the partition function is known the energy may be evaluated readily.

In quantum statistical mechanics the partition function[1] is given by—

$$Q_N = \sum_n \exp - E_n/kT \tag{2.12}$$

where E_n is the energy of the nth level. For a classical system, the analogue of equation (2.12) is evaluated from the Hamiltonian (2.1). In taking this analogue, a cell in phase space of size h^{3N} is employed because the smallest sub-division allowed by quantum mechanics is h^{3N}. It is necessary to divide by $N!$ as there are $N!$ (indistinguishable) permutations of the momenta and positions for a single phase cell. Then the resulting expression (where $\mathbf{p}_i$ is the momentum of the ith molecule) is (Huang,[3] equation 8.6)—

$$Q_N = \frac{1}{h^{3N}N!} \int_V \cdots \int e^{-H/kT}\, d(\mathbf{p}_1 \cdots \mathbf{p}_N)\, d\{N\} \qquad (2.13)$$

Employing the Hamiltonian (2.1) it is possible to perform the integration over momenta to obtain—

$$Q_N = \frac{1}{N!\Lambda^{3N}} \int_V \cdots \int \exp{-\frac{U\{N\}}{kT}}\, d\{N\} \qquad (2.14)$$

where $\Lambda = h(2\pi MkT)^{-\frac{1}{2}}$ is called the thermal wavelength of the molecule. In this form the partition function is proportional to the configuration integral Z_N (equation 2.3). Now equations (2.14) and (2.11) are combined to obtain—

$$E = \tfrac{3}{2}NkT + kT^2 \left(\frac{\partial \log(Z_N/V^N)}{\partial T}\right)_{N,V} = \tfrac{3}{2}NkT + \overline{U}_N \qquad (2.15)$$

where—

$$\overline{U}_N = \frac{\int_V \cdots \int \exp\left[-\dfrac{U\{N\}}{kT}\right] U\{N\}\, d\{N\}}{Z_N} \qquad (2.16)$$

The term $V^N/N!\Lambda^{3N}$ is the partition function for the perfect gas, and a term of this kind contributes the kinetic energy $\tfrac{3}{2}NkT$ in the above expression.

At this point a basic assumption is made upon which the whole of the later body of theory will depend: the potential energy is assumed to be equal to the sum of the potentials $[u(r)]$ developed between pairs of molecules (or atoms). In any event this term is likely to be the major term in the potential energy. The shape of $u(r)$ is shown in Figs. 3.1 and 4.1 for typical cases. Thus $U\{N\}$ is written—

$$U\{N\} = \sum_{1 \leq i \leq j \leq N} u(\mathbf{r}_{ij}) \qquad (2.17)$$

where $\mathbf{r}_{ij} = \mathbf{r}_i - \mathbf{r}_j$. Now because $u(\mathbf{r}_{ij})$ is functionally independent of the choice (i, j) there are $N(N-1)/2$ equal terms in this sum. Thus on substituting equation (2.17) into equation (2.16), the following result is obtained—

$$\overline{U}_N = \frac{N(N-1)}{2Z_N} \int_V \cdots \int \exp\left[-\frac{U\{N\}}{kT}\right] u(r_{1\,2}) \, d\{N\}$$

$$= \frac{1}{2} \iint_V u(r_{1\,2}) n_N^{(2)}(r_{1\,2}) \, dr_1 \, dr_2$$

and allowing $N \to \infty$

$$\left\{\frac{\overline{U}_N}{N}\right\}_{N \to \infty} = \frac{\rho}{2} \int_V u(r) g(r) \, dr \tag{2.18}$$

where $\rho = N/V$ is the number density. The formula for the total internal energy is obtained by using this result in equation (2.15)

The result (equation 2.18) may be obtained through a descriptive argument as follows—

   (i) the average number of molecules at a distance between r and $r+dr$ from a given molecule is $(4\pi r^2 \, dr)\rho g(r)$;

   (ii) the average potential energy of interaction with these neighbours is $(4\pi r^2 \, dr)\rho g(r) u(r)$;

   (iii) integrate over r to get the total potential energy;

   (iv) this result is for one molecule, so multiply by N for the whole system and divide by 2 to avoid counting each pair interaction twice;

   (v) the result is $(N/2)\rho \int u(r)g(r)4\pi r^2 \, dr$ as in equation (2.18).

Equation (2.15) may be used in standard thermodynamic formulae to derive other quantities. For example, the specific heat (per molecule) at constant volume is given by [1]—

$$C_V = \frac{1}{N}\left(\frac{\partial E}{\partial T}\right)_V = \frac{3}{2}k + \frac{\rho}{2} \int \left(\frac{\partial g(r)}{\partial T}\right)_V u(r) \, dr \tag{2.19}$$

It was assumed in equation (2.17) that $u(r)$ is independent of temperature so that only the temperature dependence of $g(r)$ appears in equation (2.19). Clearly a precise knowledge of the variation of $g(r)$ with T at constant density is required before equation (2.19) is useful. It may be seen from Table 1.1 line 7 that $\rho/2 \int (\partial g/\partial T)_V u(r) \, dr$ is of the order of k at the triple point.

2.4. The Equation of State

The relationship between pressure, temperature and volume may be calculated by applying the above methods to the usual thermodynamic expressions (Landau and Lifshitz,[1] equations 15.4 and 31.3)—

$$P = -\left(\frac{\partial F}{\partial V}\right)_T = kT\left(\frac{\partial \log Q_N}{\partial V}\right)_{N,T} \qquad (2.20)$$

where F is the free energy. Equation (2.14) for the partition function involves an integration over volume. In order to make U an explicit function of V, the limits of integration in equation (2.14) are changed by the substitution—

$$\mathbf{r} = \mathbf{r}'\sqrt[3]{V} \qquad (2.21)$$

Hence—

$$Q_N = \frac{V^N}{N!\Lambda^{3N}}\int_0^1 \cdots \int_0^1 \exp -\frac{U\{N\}}{kT}\,\mathrm{d}\{N\} \qquad (2.22)$$

The factor V^N gives a term $(N/V)kT$ when equation (2.22) is used in equation (2.20). The remaining factor involves the differential $\partial U/\partial V$, which by virtue of equation (2.17) may be written—

$$\frac{\partial U}{\partial V} = \sum_{1 \le i \le j \le N} \frac{\mathrm{d}u(r_{ij})}{\mathrm{d}(r_{ij})} \frac{r_{ij}}{3V} \qquad (2.23)$$

The next step is to reverse the transformation (equation 2.21) and complete the calculation by the method used to calculate $\bar{U}$ in the previous Section. Thus—

$$P = \frac{N}{V}kT - \frac{1}{Z_N}\int_V \cdots \int \exp\left[-\frac{U\{N\}}{kT}\right]\frac{\partial U}{\partial V}\,\mathrm{d}\{N\}$$

$$= \frac{N}{V}kT - \frac{\rho}{6}\iint_V r_{12}\frac{\mathrm{d}u(r_{12})}{\mathrm{d}r}n_N^{(2)}(r_{12})\,\mathrm{d}\mathbf{r}_1\,\mathrm{d}\mathbf{r}_2$$

on taking the $N \to \infty$ limit—

$$= \frac{N}{V}kT\left[1 - \frac{\rho}{6kT}\int_V r\frac{\mathrm{d}u(r)}{\mathrm{d}r}g(r)\,\mathrm{d}\mathbf{r}\right] \qquad (2.24)$$

This equation of state has been calculated within the framework of the canonical ensemble. It is usually termed the "pressure equation", or the "virial equation" as it may be derived from the virial theorem of classical mechanics.

Another (but equivalent) equation is obtained by working with the grand canonical ensemble, and considering the fluctuations in number molecules in a given volume. In this case the pressure is given by the generalization of equation (2.20) (Huang,[3] equation (8.35), differentiating with respect to V and employing the result $(\partial P/\partial V)_{\mu,T} = 0$) i.e.—

$$P = kT \left(\frac{\partial \log \mathscr{L}}{\partial V} \right)_{\mu,T} \tag{2.25}$$

where $\mathscr{L}$ is the grand partition function defined by equation (2.7). This equation is differentiated with respect to the chemical potential μ, to obtain—

$$\left[\frac{\partial}{\partial V} \left(\frac{\partial \log \mathscr{L}}{\partial \log z} \right)_{V} \right]_{\mu,T} = \left(\frac{\partial P}{\partial \mu} \right)_{V,T} \tag{2.26}$$

Then the righthand side is rewritten as—

$$\left(\frac{\partial P}{\partial \mu} \right)_{V,T} = \left(\frac{\partial P}{\partial \rho} \right)_{V,T} \left(\frac{\partial \rho}{\partial \mu} \right)_{V,T} = \frac{1}{V} \left(\frac{\partial P}{\partial \rho} \right)_{V,T} \left(\frac{\partial \bar{N}}{\partial \mu} \right)_{V,T} \tag{2.27}$$

since $\rho = \langle N \rangle / V$. Now from the definition of $\mathscr{L}$ it can be shown that—

$$\bar{N} \equiv \langle N \rangle = \frac{1}{\mathscr{L}} \sum N \frac{z^N Z_N}{N! \Lambda^{3N}} = \frac{\partial \log \mathscr{L}}{\partial \log z}$$

and—

$$\bar{N}^2 \equiv \langle N^2 \rangle = \frac{1}{\mathscr{L}} \sum N^2 \frac{z^N Z_N}{N! \Lambda^{3N}} = \frac{1}{\mathscr{L}} \frac{\partial^2 \mathscr{L}}{\partial (\log z)^2}$$

so that—

$$\left(\frac{\partial \langle N \rangle}{\partial \mu} \right)_{V,T} = \frac{\langle N^2 \rangle - \langle N \rangle^2}{kT}$$

Thus equations (2.26) and (2.27) may be written—

$$\rho = \frac{1}{VkT} \left(\frac{\partial P}{\partial \rho} \right)_{V,T} [\langle N^2 \rangle - \langle N \rangle^2] \tag{2.28}$$

Finally through the use of equation (2.9) with equation (2.8), equation (2.28) becomes [since $\bar{N}^2 - (\bar{N})^2 = (\bar{N}^2 - \bar{N}) - (\bar{N})^2 + \bar{N}$]—

$$\left(\frac{\partial \rho}{\partial P} \right)_{V,T} kT = \rho \int_V (g(r) - 1) \, dr + 1 \tag{2.29}$$

This result connects pressure, temperature and density and so defines an equation of state. The left-hand side is equal to $\rho\chi_T \cdot kT$ and, for this reason equation (2.29) is known as the "compressibility" equation.

In a later Chapter both equations (2.24) and (2.29) will be used to make numerical estimates of the equation of state, and by comparison of the two results the accuracy of the estimates may be tested.

2.5. Cluster and Virial Expansions of the Equation of State

Consider the function Z_N—equation (2.3)—when the pair potential approximation (2.17) is used for $U\{N\}$. In this approximation, equation (2.3) becomes—

$$
\begin{aligned}
Z_N &= \int_V \cdots \int d\{N\} \prod_{i<j} \exp -\frac{u(r_{ij})}{kT} \\
&= \int_V \cdots \int d\{N\} \prod_{i<j} (1+f_{ij}) \\
&= \int_V \cdots \int d\{N\}[1+(f_{12}+f_{13}\ldots)+(f_{12}f_{13}+f_{12}f_{14}\ldots)+\text{etc.}]
\end{aligned}
$$

(2.30)

where $\exp[-u(r_{ij})/kT] \equiv 1+f_{ij}$. This expansion is convenient because f_{ij} is a bounded function; that is when r_{ij} is small $f_{ij} \to 1$ even though the repulsive forces between the atoms cause $u(r)$ to become large and positive. Also note that in the expansion (equation 2.30) the first term involves single molecules, the second pairs of molecules the third triplets and so on. For this reason it is known as a cluster expansion, and since many properties are related to Z_N it is possible to obtain cluster expansions for them too.

Since both the grand partition function (equation 2.7) and the partition function (equation 2.14) are given in terms of Z_N an expansion of the equation of state (equations 2.20 and 2.25) in this form is possible. Because the algebraic work is rather long, it will not be reproduced here; only the result (Huang,[3] equation 14.28) will be quoted—

$$
\left.
\begin{aligned}
\frac{P}{kT} &= \frac{1}{\Lambda^3} \sum_{l=1}^{\infty} b_l z^l \\
\frac{N}{V} &= \frac{1}{\Lambda^3} \sum_{l=1}^{\infty} l b_l z^l
\end{aligned}
\right\}
$$

(2.31)

where the coefficients b_l involve the f_{ij}. If the system is very dilute, an expansion in powers of $1/V$ is useful. This expansion is called the virial expansion and is defined as—

$$\frac{P}{kT} = \frac{N}{V} \sum_{l=1}^{\infty} a_l(T) \left(\frac{N}{V}\right)^{l-1}$$ (2.32)

The relationship between the coefficients in these two equations is obtained by equating the right-hand sides of equations (2.31) and (2.32) after taking the $V \rightarrow \infty$ limit of the b_l.

The first two terms in equation (2.32) may be evaluated from equation (2.24) by making the dilute gas approximation for $g(r)$—

$$g(r) = \exp-\frac{u(r)}{kT}$$ (2.33)

Thus from equation (2.24)—

$$P = \frac{NkT}{V}\left[1 - \frac{\rho}{6kT}\int_0^{\infty} 4\pi r^3 \, dr \, \frac{du}{dr} \exp-\frac{u(r)}{kT}\right]$$ (2.34)

The integral in this expression can be written in the form—

$$kT\int_0^{\infty} 4\pi r^3 \, d\left(1 - \exp-\frac{u(r)}{kT}\right)$$

$$= \left[r^3\left(1 - \exp-\frac{u(r)}{kT}\right)\right]_0^{\infty} - 12\pi kT \int_0^{\infty} r^2 \left(1 - \exp-\frac{u(r)}{kT}\right) dr$$

and hence equation (2.34) becomes—

$$\frac{P}{kT} = \rho + 2\pi\rho^2 \int_0^{\infty} r^2\left(1 - \exp-\frac{u(r)}{kT}\right) dr$$ (2.35)

By comparison with equation (2.32), the first two coefficients are seen to be—

$$\left.\begin{aligned} a_1 &= 1 \\ a_2 &= -2\pi\int_0^{\infty} r^2 f_{1\,2}(r) \, dr \end{aligned}\right\}$$ (2.36)

Equation (2.36) is sometimes used to obtain data on $u(r)$ from experimental measurements of the equation of state for a gas. This application is discussed in Section 3.8. Many of the higher terms in the expansion (2.32) have been worked out and are given in the literature.[5]

In Chapter 5 a cluster expansion will be used to discuss the relations between $g(r)$ and $u(r)$ and expressions for some of the higher-order terms will be given (equation 5.12). These expansions are very useful for discussing systems at low density, but become difficult to use if many terms are employed. Thus in the high-density case it is normal to make an approximation to the cluster series which allows the density expansion to be summed so giving a closed expression. The validity of such results depends upon the physical situation; for example if there is a smooth variation of a property from low density to high density, the use of this technique can be justified. If however there is an abrupt change in the property, at a phase change for example, then it is unlikely that this method will be satisfactory.

Finally the pair potential approximation (equation 2.17) should be emphasized as the basic approximation made here. If this approximation is accepted a logical discussion of liquid-state properties can be given; some properties have been discussed in this Chapter and others are to be examined in succeeding Chapters.

Symbols for Chapter 2

a_l	Coefficient in virial expansion	E_n	Energy of nth energy level
b_l	Coefficient in cluster expansion	Λ	Thermal wavelength of molecule

CHAPTER 3

The Pair Potential Function for Non-Conducting Liquids

In Chapter 2 it was shown how several important properties may be calculated on the basis of pair theory provided the functions $g(r)$ and $u(r)$ are known. The discussion of the uses of these theoretical results is deferred until Chapter 7, so that the theoretical and experimental methods of studying $g(r)$ and $u(r)$ may be examined. This part of the discussion will be divided into three parts: (i) theoretical and experimental studies of $u(r)$ (Chapters 3 and 4); (ii) theoretical relationships between $g(r)$ and $u(r)$ (Chapter 5); and (iii) experimental methods of studying $g(r)$ (Chapter 6). The pair potential $u(r)$ differs markedly between non-conducting (electrical) and conducting liquids, as indicated in Chapter 1, and for this reason separate treatments of these two cases will be given. If the pair potential approximation was a good one, $u(r)$ would be independent of external variables (volume, temperature, etc.) but this can never be completely true. In the case of metals (Chapter 4), many body forces are included in the definition of $u(r)$, whereas for non-conducting liquids this is not believed to be necessary. Because of this, the "effective" $u(r)$ for a metal will vary more with (e.g.) volume than the "real" $u(r)$ for a non-conducting liquid. Consequently, the pair theory of liquids should be looked as a double theory including (a) the "perfect" pair potential theory in which many body forces are negligible and (b) the "effective" pair potential theory in which the effects of many body forces are included in the parameters describing the pair potential.

3.1. A General Restriction on $u(r)$

The pair approximation itself (equation 2.17) imposes a restriction upon the form of $u(r)$ because the assumption behind equation (2.17) is that the potential energy of a thermodynamically stable system can be described as a sum of pair potentials, $u(\mathbf{r}_{ij})$. Clearly some limits are placed upon $u(r)$ by the requirement that the system is thermodynamically stable. Another way of expressing this point is to say that the grand partition function—equations (2.7) and (2.3)—must be convergent, and this requires that—

$$U\{N\} = \sum_{1 \le i \le j \le N} u(\mathbf{r}_i - \mathbf{r}_j) \ge -NB \qquad (3.1)$$

where $B \ge 0$ is a positive constant. One set of conditions for equation (3.1) to be satisfied, when working in a three-dimensional space and when $u(r)$ is bounded from below, is[9]—

$$\left.\begin{array}{ll} u(r) > C_1/r^{3+\delta} & \text{for } r \to 0 \\ |u(r)| < C_2/r^{3+\delta} & \text{for } r \to \infty \end{array}\right\} \qquad (3.2)$$

where C_1, C_2 and δ are positive constants. The first of these conditions ensures that the core of the potential is sufficiently repulsive that the system does not collapse while the second ensures that the potential is sufficiently short ranged that the partition functions remain bounded. In practice with the potentials discussed below, equation (3.2) will be easily satisfied. Sections 3.2 to 3.4 are concerned with non-conducting liquids and Chapter 4 will deal with liquid metals.

3.2. Dipolar Attraction

The electrical attraction between dipoles occurs both for those molecules possessing an electric dipole moment and also for atoms possessing a spherically symmetric change distribution. In the latter case, the orbital motion of the electrons causes (at any given instant) the electron shells of an atom to be displaced relative to the nucleus, or to be distorted in shape, producing an instantaneous dipole: this dipole induces a dipole moment on any other atom so that there is a mutual attraction. This effect gives rise to an attractive force called the van der Waals force. It is attractive and temperature independent to a good approximation so that it meets the assumed properties of $u(r)$.

The variation of the van der Waals force with separation (r) between the atoms can be calculated by noting that the electric field (E), at a large distance r from the centre of the dipole, varies as $1/r^3$. Now the force (F) on a dielectric placed in a field E is given by—

$$F \propto -\frac{d}{dr}(E^2) \propto \frac{1}{r^7} \qquad (3.3a)$$

This force is equivalent to a potential varying as—

$$u(r) \propto -\frac{1}{r^6} \quad \text{for large } r \quad (3.3b)$$

and the constant of proportionality is related to the ease of forming dipoles, i.e. the polarizability of the atom.

The same result may be obtained by a quantum mechanical calculation. If there are two nuclei at a separation R with n electrons and V is the sum of the electrostatic interactions between all pairs of them, then the Schrödinger equation may be written in the form—

$$\left\{ -\frac{\hbar^2}{2m_e} \sum_{i=1}^{n} \nabla_i^2 + V \right\} \phi_R\{\mathbf{r}_j\} = U(\mathbf{R})\phi_R\{\mathbf{r}_j\} \quad (3.4)$$

where $\phi(\mathbf{r}_j)$ is the electronic wave function of the jth electron and $U(R)$ is the nuclear potential function (i.e. $U(R)$ is an eigenvalue of equation 3.4). If, for example, there are only two electrons (as for two hydrogen atoms) then V has the form—

$$V = e^2 \left[\frac{1}{R} + \frac{1}{r_{12}} - \frac{1}{r_{1A}} - \frac{1}{r_{2B}} - \frac{1}{r_{1B}} - \frac{1}{r_{2A}} \right] \quad (3.5)$$

where the suffixes $(1, 2)$ denote the electrons and (A, B) denote the nuclei. The attractions between atoms are developed by all terms in equation (3.5) connecting the two atoms, i.e. all except terms (1A) and (2B). Thus the Hamiltonian for this problem may be written as—

$$H = H_0 + H'$$

where H_0 is the unperturbed Hamiltonian and the perturbation caused by the interaction between the atoms is—

$$H' = e^2 \left[\frac{1}{R} + \frac{1}{r_{12}} - \frac{1}{r_{1B}} - \frac{1}{r_{2A}} \right]$$

$$= a_{22}\frac{e^2}{R^3} + a_{24}\frac{e^2}{R^4} + a_{44}\frac{e^2}{R^5} + \text{etc.} \quad (3.6)$$

where the a's are appropriate coefficients in the expansion of H' in powers of $1/R$ (see Hirschfelder et al.[10] for details of this expansion). These three terms are, respectively, the equivalent of dipole–dipole, dipole–quadrupole and quadrupole–quadrupole interactions. Finally the interaction energy is proportional to H'^2 and hence its leading term varies as $1/R^6$. This argument may be extended to other atoms containing larger numbers of electrons, provided the pair of atoms are in non-degenerate spherically symmetric ground states. It may be seen from the above that this interaction is temperature independent provided the electronic orbits are temperature independent.

3.3. Repulsive Terms

The repulsion between two spherically symmetric atoms arises when they are sufficiently close together that their electron shells overlap. A detailed quantum mechanical calculation of this interaction is complicated and leads to a potential which is mathematically inconvenient. However the result may be expressed in the form—

$$u(r) \simeq \sum_i P_i(r) \exp\left(-\frac{r}{r_{ci}}\right) \qquad (3.7)$$

where r_{ci} are constants and P_i are various polynominals in r. Now it is found that the r_{ci} are roughly equal and that $P_i(r)$ is roughly constant over a useful range of r. In this context the useful range of r is that for which equation (3.7) is about one to three times the absolute value of the attractive potential (Section 3.4). This is the only region of importance for atoms of thermal energies. Thus in this region it is customary to write—

$$u(r) \sim \text{constant. } \exp(-r/r_c) \qquad (3.8)$$

and to set $u(r) \to \infty$ for values of r less than an arbitrary small value. Sometimes, for mathematical convenience, equation (3.7) is replaced by a power law, i.e.—

$$u(r) \sim \frac{\text{constant}}{r^n} \qquad (3.9)$$

where the value of n is chosen to give the best fit over the useful range of r (see Hirschfelder *et. al.*[10] for a detailed discussion of these points).

3.4. Some Convenient Potentials

In this Section three potentials often used in numerical calculations are listed—

(A) the Lennard–Jones potential—

$$u(r) = 4\varepsilon\left[\left(\frac{\sigma}{r}\right)^{12} - \left(\frac{\sigma}{r}\right)^6\right] \qquad (3.10)$$

(B) the modified Buckingham potential—

$$u(r) = \frac{\varepsilon}{1-6/\alpha}\left[\frac{6}{\alpha}\exp\alpha\left(1-\frac{r}{r_m}\right) - \left(\frac{r_m}{r}\right)^6\right]_{r>r_{max}} \qquad (3.11)$$

$$= +\infty \quad \text{for } r \leq r_{max} \text{ (See Fig. 3.1)},$$

(C) the hard sphere potential—

$$u(r) = +\infty \quad \text{for } r < \sigma \left.\right\}$$
$$\quad\;\; = 0 \quad\;\;\; \text{for } r \geq \sigma \left.\right\}$$

(3.12)

These potentials are illustrated in Fig. 3.1. In each of cases (A) and (B), the parameter ε is the depth of the attractive well, and the long-range attraction is given by the dipole–dipole term. They differ in the

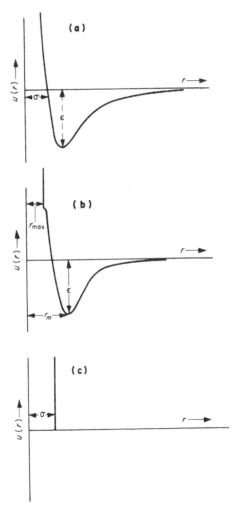

FIG. 3.1. Illustrations of the spherically symmetric potentials given in equations (3.10) to (3.12); (a), L–J; (b), modified Buckingham; (c), hard sphere.

choice of the representation for the repulsive term (the main term in equation 3.11 has a maximum at $r = r_{max}$, and u is put equal to ∞ there). The potential (C) is often convenient for mathematical reasons, as will be seen later. However, it can only be used for those problems in which the hard core of the potential is the important feature. It is notable that all three potentials satisfy the restrictions of equation (3.2). In order to use these expressions, it is necessary first to confirm experimentally the analytical form and secondly to measure the values of the constants. The methods of doing this will be described now.

3.5. Classical Atom–Atom Scattering Experiments

The theoretical outline described in Sections 3.2 to 3.4 is applicable to atoms or molecules which are electrically neutral, preferably without an electric moment and of spherical shape. Such systems (e.g. rare-gas atoms, globular molecules like CH_4, SF_6, etc.) should possess a $u(r)$ which is independent of density, and in particular can be studied by measuring the interaction (i.e. scattering) of an isolated pair of atoms. Throughout this Chapter only the *elastic* scattering is considered, that is processes which do not alter the quantum state of the colliding atoms.

The calculation (Kennard,[11] Chapter III) of the classical scattering cross-section for neutral atoms starts from a consideration of the *relative** motion as shown in Fig. 3.2. An atom of reduced mass ($m = m_1 m_2/(m_1 + m_2) = \frac{1}{2}$ actual mass for equal atoms) is approaching another which is placed at the origin O. The perpendicular distance (b) from O to the initial path is called the impact parameter, and the scattering angle (θ) is to be evaluated in terms of this parameter. For convenience the angle θ_1, defined in Fig. 3.2 will be used. The angular momentum of the first atom about the origin is $mr^2\dot{\theta}_1$ and its kinetic energy is $\frac{1}{2}m(\dot{r}^2 + r^2\dot{\theta}_1^2)$, where r is its distance from O and dots denote differentiation with respect to time. Since the initial values of these quantities are mvb and $\frac{1}{2}mv^2$, the equations of motion are—

$$mvb = mr^2\dot{\theta}_1; \qquad \tfrac{1}{2}mv^2 = \tfrac{1}{2}m(\dot{r}^2 + r^2\dot{\theta}_1^2) + u(r) \qquad (3.13)$$

These equations are solved to give $\dot{\theta}_1$ and $\dot{r}$ and hence the following relationship—

$$\frac{d\theta_1}{dr} = \frac{\dot{\theta}_1}{\dot{r}} = \frac{b}{r^2}\left[1 - \frac{u(r)}{E} - \frac{b^2}{r^2}\right]^{-\frac{1}{2}} \qquad (3.14)$$

where the energy E has been written in place of $\frac{1}{2}mv^2$.

---

* That is, the velocity v is the relative velocity and the origin is placed on the target atom; this is the simplest frame in which to do the calculation.

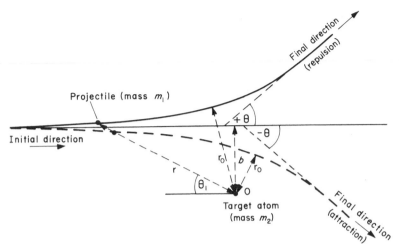

FIG. 3.2. Geometry of atom–atom scattering experiment; paths and parameters for both repulsive (full line) and attractive (dashed line) potentials are shown.

Because the path is symmetrical about its point of closest approach to O (distance r_0), the total increase in θ_1 during the collision process is—

$$2\int_{r_0}^{\infty} \left| \frac{d\theta_1}{dr} \right| dr$$

if the positive root is taken in equation (3.14). From Fig. 3.2 this change in θ_1 can be seen to be equal to $\pi - \theta$, thus—

$$\theta(b) = \pi - 2b \int_{r_0}^{\infty} r^{-2} \left[1 - \frac{u(r)}{E} - \frac{b^2}{r^2} \right]^{-\frac{1}{2}} dr \qquad (3.15)$$

The value of r_0 is calculated from the observation that when $r = r_0$, the tangent to the path is normal to r_0 and thus $\dot{r} = 0$. Using this condition in (3.13) gives—

$$r_0 = b \left[1 - \frac{u(r_0)}{E} \right]^{-\frac{1}{2}} \qquad (3.16)$$

If the scattering cross-section is $\sigma(\theta)$, the fraction of atoms scattered into the range $d\theta$ is $2\pi\sigma(\theta) \sin \theta \, d\theta$. But this is equal to the fraction of atoms in the incident beam which have impact parameters between b and $b + db$, where $db = d\theta/|d\theta/db|$. Since a plane containing all such impact parameters is normal to the beam, it follows that—

$$2\pi b \, db = 2\pi\sigma(\theta) \sin \theta \, d\theta$$

or—

$$\sigma(\theta, E) = \frac{b}{\sin \theta} \left| \frac{db}{d\theta} \right| \qquad (3.17)$$

Hence the cross-section is readily calculated from equations (3.15) and (3.17) if $u(r)$ is known, or conversely if the cross-section as a function of E and θ is measured, $u(r)$ may be determined by numerical analysis.

The above calculation is purely classical and hence applies for those cases where r is large compared to the atomic wavelength. The most interesting case is the limit $r \gg \sigma$ (Fig. 3.1) where it is implied that $u(r)$ is small since it was shown in Section 3.2 that $u(r)$ falls off rapidly with r. Thus in this limit, a simplification can be made by expanding equations (3.15) and (3.16) in powers of $u(r)/E$ and $u(r_0)/E$, respectively. Truncation to first order and combination of the results yields—

$$\theta(b) \sim \frac{1}{E} \int_0^1 \frac{u(r_0) - u(r_0/x)}{(1 - x^2)^{\frac{3}{2}}} \, dx + \text{etc.} \tag{3.18a}$$

where $x = r_0/r$, and from equation (3.16)—

$$r_0 \sim b \left(1 + \frac{u(r_0)}{2E} + \text{etc.} \right) \tag{3.18b}$$

Equation (3.18) demonstrates that in this approximation the angle of scatter will be small. In order to proceed further it is necessary to assume a form for $u(r)$. In line with Section 3.2, assume $u(r) = -C/r^6$ and evaluate equation (3.18) to first order, giving—

$$\theta(b) \sim \frac{15\pi c}{16 E r_0^6} \simeq \frac{15\pi c}{16 E b^6} + O\left(\frac{u(r_0)}{E}\right) \tag{3.19}$$

Finally, by using equation (3.17), the cross-section in this approximation is—

$$\sigma(\theta, E) \sim \frac{b^2}{6\theta^2} = \frac{1}{6}\left(\frac{15}{16}\pi c\right)^{\frac{1}{3}} \frac{1}{E^{\frac{1}{3}}} \cdot \frac{1}{|\theta|^{\frac{7}{3}}} \tag{3.20}$$

Thus equation (3.20) provides a means of testing atomic interactions for the existence of the long-range dipolar potential, or if written in the form $\theta^{-[(2s+1)/s]}$ for any long-range power law (r^{-s}) potential. It is of limited application, however, and must be supplemented by the full classical equation and by quantum mechanical results. For example, equation (3.20) shows that $\sigma(\theta)$ diverges to infinity as $\theta \to 0$, because the potential falls off slowly and there is always a weak interaction even at large distances. From the classical point of view, a stationary atom placed in a wide parallel beam of atoms would throw out a divergent shadow (rather than a parallel shadow) and so give rise to an infinite cross-section. But quantum mechanical diffraction effects will restrict the size of the shadow and cause the cross-section to be finite (on this picture the classical theory will fail for $\theta \sim \lambda/2r_0 = \pi h/mvr_0$ where λ is the wavelength of the projectile). Even in the quantum mechanical case there is a large amount of small angle scattering which dominates both the experimental and theoretical discussion of the total cross-section.

3.6. Quantum Mechanical Calculation of Atom–Atom Scattering

In quantum mechanics[6,7] it is necessary to solve the wave equation for the relative motion, i.e.—

$$-\frac{\hbar^2}{2m}\nabla^2\phi + u(r)\phi = E\phi \tag{3.21}$$

where m is the reduced mass and ϕ is the required wave function. In solving this equation, only elastic scattering is considered (i.e. processes in which the internal states of the atoms are unchanged) since such processes are the simplest ones which contain all the information on $u(r)$.

An outline only (following Schiff,[6] Chapter V) will be given of how this equation is solved for the cross-section for scattering by a spherically symmetric potential. Since the scattered particles are observed at large distances from the scattering centre, only the asymptotic form of ϕ is required. This is the sum of an incident and a scattered wave, i.e.—

$$[\phi(r, \theta)]_{r \to \infty} \to \text{Constant}\left[e^{iqz} + \frac{f(\theta)}{r}e^{iqr}\right] \tag{3.22}$$

where $q = mv/\hbar$, and the required cross section is simply—

$$\sigma(\theta) = |f(\theta)|^2 \tag{3.23}$$

The angular and radial parts of the wave function are separated by the substitution—

$$\phi(r, \theta) = \sum_{l=0}^{\infty} \frac{\chi_l(r)}{r} P_l(\cos\theta) \tag{3.24}$$

where P_l is the lth Legendre polynomial. In this case, equation (3.21) reduces to a set of equations for the χ_l, i.e.—

$$\frac{\hbar^2}{2m}\frac{d^2\chi_l}{dr^2} + \left[1 - \frac{u(r)}{E} - \frac{b^2}{r^2}\right]E\chi_l = 0 \tag{3.25}$$

where $b = \sqrt{l(l+1)}/q$ is the analogue of the impact parameter. This follows from the fact that the angular momentum is $\hbar\sqrt{l(l+1)}$ which is equal to mvb in the classical case. It is interesting to note that the quantities in the square brackets of equations (3.25) and (3.14) are the same.

According to equation (3.22), the asymptotic form of χ_l is required, and this may be shown[6,7] to have the form—

$$[\chi_l(r)]_{r\to\infty} \to \frac{2l+1}{q} i^l \exp(i\delta_l) \sin(qr - \frac{l\pi}{2} + \delta_l) \qquad (3.26)$$

The angle δ_l is called a "phase shift", since it is the phase difference between the asymptotic form of the actual radial wave function and the radial wave function in the absence of the potential $u(r)$. Thus the δ_l now contain the information concerning $u(r)$. Equation (3.26) may be used in equations (3.24) and (3.22)* to evaluate $f(\theta)$ as—

$$f(\theta) = \frac{1}{2iq} \sum_{l=0}^{\infty} (2l+1)[\exp(2i\delta_l)-1] P_l(\cos\theta) \qquad (3.27)$$

and the scattering cross-section follows immediately from equation (3.23). In practice, the observed cross-section—now an oscillatory function of θ, $\sigma(\theta)$ at a given energy E—would be fitted by the partial wave series (equation 3.27), so obtaining experimental values of δ_l as a function of E. The validity of a particular potential may be examined by calculating the phase shifts from equation (3.25), and then comparing with experiment.

In the limiting case when the scattered wave can be treated as a small perturbation the value of the phase shift may be obtained directly in terms of $u(r)$. This result may be shown (Schiff,[6] equation 26.27) to be—

$$\delta_l \sim -\frac{2m}{\hbar^2} \int_0^{\infty} qr^2 j_l^2(qr)u(r)\,dr \qquad (3.28)$$

where j_l is the lth spherical Bessel function. Also, in this case the phase shifts are small so that $\exp(2i\delta_l)-1 \sim 2i\delta_l$, and $f(\theta)$ becomes—

$$f(\theta) \sim \frac{1}{q} \sum_{l=0}^{\infty} (2l+1)\delta_l P_l(\cos\theta) \qquad (3.29)$$

Taking equations (3.28) and (3.29) together gives (if $Q = 2q\sin\theta/2$)—

$$f(\theta) = -\frac{1}{4\pi} \int_V \exp(i\mathbf{Q}\cdot\mathbf{r})\, u(r)\,d\mathbf{r} \qquad (3.30)$$

* Exp(iqz) is expanded in Legendre polynominals also.

where the relation—

$$\frac{\sin \mathbf{Q} \cdot \mathbf{r}}{\mathbf{Q} \cdot \mathbf{r}} = \sum_{l=0}^{\infty} (2l+1) j_l^2(qr) P_l(\cos \theta)$$

has been used. Equation (3.30) shows that in this limit (known as the Born approximation) the angular distribution is related to the Fourier transformation of the scattering potential.

It may be shown[7] that the phases will be small (for $qr \sim l+\frac{1}{2}$) if—

$$\frac{2m}{\hbar^2} u(r) \ll \frac{l(l+1)}{r^2} \quad \text{or} \quad \frac{u(r)}{E} \ll \frac{b^2}{r^2}$$

Clearly for large r this is always so, since by equation (3.2) the function $u(r)$ falls off faster than r^{-2}. Thus the large r contribution to the cross-section (which gave an infinite contribution in the classical calculation) may be calculated from the Born approximation. If $u(r)$ vanishes faster than r^{-3}, equation (3.30) will give a finite cross-section[7] for all θ, and the total cross-section is found to be proportional to $q^{-\frac{2}{5}}$ (or $E^{-\frac{1}{5}}$) when $u(r)$ varies as r^{-6}. In the general case (potential r^{-s}) the total cross-section is found to vary as $q^{-[2/(s-1)]}$.

Three methods of treating atom–atom scattering have been outlined above: the classical method; the full quantum mechanical partial wave treatment; and the Born approximation. No one of these methods is satisfactory for analysing the experimental data, because the full method is very difficult to handle and the range of the limiting methods is not very great. Some examples will be given in the next Section to illustrate how the forms for $u(r)$ can be tested.

It is useful to compare the conditions for the validity of the classical and Born approximations. These conditions are summarized in the following Table[7]—

Classical Approximation	Born Approximation
$V\sigma/\hbar V \gg 1$	$V\sigma/\hbar V \ll 1$, for $q\sigma \gg 1$.
δ_l and l large	$\delta_l \to 0$ for all l.

where σ, V are the order of magnitude of the range and depth of the potential. Thus although these approximations may have appeared to be superficially similar, they represent opposite extremes.

3.7. Experimental Test of Small (r) and Large (r) Form for $u(r)$

The most convenient tests are to compare the total cross-section to the quantum mechanical formula and the angular distribution data to the classical formula. Two series of experiments will be mentioned,[12] in which neutral atoms were scattered from one another. In such experiments it is frequently convenient to use metal atoms for either the projectile or the target (or both) since atomic beams of metal atoms are convenient to prepare. As a test of dipolar attraction this is satisfactory because the beams consist of single neutral atoms. Figures 3.3 (a), (b) and (c) show the comparison with (i) the total cross-section $v^{-\frac{2}{5}}$ prediction and (ii) the angular distribution prediction $\theta^{-\frac{7}{3}}$ at thermal velocities. In both cases reasonable agreement is found between the theoretical result and the mean cross-section confirming that the potential between neutral atoms varies as r^{-6} at long range.

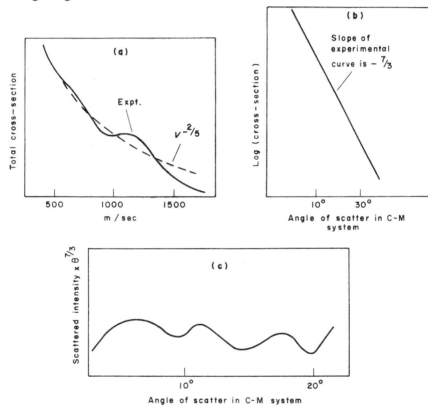

FIG. 3.3. Examples of data from atom–atom scattering experiments, illustrating the r^{-6} term in the inter-atomic potential. The oscillations about the classical curve due to quantum mechanical effects are clearly seen: (a), K atoms on Kr, from E. Rothe (see ref. 12, p. 927); (b), 439 m/sec K atoms on Hg, from R. B. Bernstein (see ref. 12, p. 895); (c), [7]Li atoms on Hg, from R. B. Bernstein (see ref. 12, p. 895).

Figure 3.4 shows the variation of $u(r)$ with r deduced by Amdur (see reference 12, p. 934) from a series of total-cross-section experiments with fast neutral atoms (500–2000 eV). Because of the high energy, the repulsive part of the potential is observed. Each section of the cross-section versus energy curve was fitted by the formula—

$$\sigma_T(v) \sim v^{-[2/(s+1)]} \quad \text{for} \quad u(r) \sim 1/r^s \qquad (3.31)$$

in order to find the value of s for a given range of r (given by $r \sim h/mv$). In this way, a composite curve was built up of the variation of $u(r)$ with r, and is shown in Fig. 3.4. These data confirm that the repulsive term is roughly exponential, and differs significantly from the

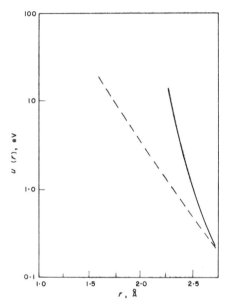

FIG. 3.4. Experimental test of repulsive part of $u(r)$ from argon–argon scattering, from I. Amdur (see ref. 12, p. 934). The dashed line is the mean of the experimental results, and since it is nearly linear on this graph, it is roughly exponential, as expected theoretically. The full line is the form of an L–J potential normalized to the experiment at $r = 2.75\,\text{Å}$.

Lennard–Jones approximation. However since the exact form of the potential at such high bombarding energies is not very significant to liquid-state problems, the broad general agreement between the Buckingham or L–J potential and experiment shown in Figs. 3.3 and 3.4 can be considered satisfactory (especially since the Buckingham potential fails badly at very small r—Fig. 3.1).

3.8. Evaluation of the Constants of the Model Potentials

Ideally the constants in the models for $u(r)$ would be evaluated by fitting the model to the result of a quantum mechanical calculation of the force between two atoms. The data obtained by this method are in reasonable agreement with those obtained by experiment,[10] as shown at Fig. 3.5 for neon. However it is perhaps more direct to fit the model to the physical properties of the gas, and then try to calculate liquid properties using that potential. In this way, the liquid

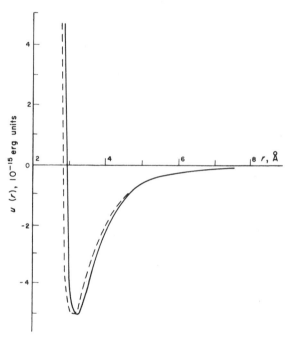

FIG. 3.5. Comparison of the basic calculation of $u(r)$ for neon with the L–J model: full line, calculated by Bleick and Meyer, *J. Chem. Phys.*, 1933, **44**, 214; dashed line, L–J potential using parameters derived from gas viscosity. (From Hirschfelder *et al.*,[10] Fig. 14.2.2, p. 1072.)

properties are predicted in terms of the experimental gaseous properties, and quantum mechanics is used to calculate the general form of the potential.

Any property dependent upon molecular collisions may be used to evaluate the parameters of $u(r)$. Those commonly employed are the virial coefficients and the gaseous transport coefficients, which depend ultimately upon the atom–atom scattering cross-sections discussed in Sections 3.5 and 3.6. The virial coefficients are determined experimentally by fitting the equation-of-state data to the series (equation 2.32). Of particular value is the second virial coefficient (equation 2.36) which has been measured accurately as a function of temperature. Table 3.1 gives the results[10] obtained by using equations (3.10) and (3.11) in equation (2.36) and fitting to these data. There is general agreement between the depth and range of the potential deduced on the basis of these two models.

TABLE 3.1
Model parameters obtained from second virial coefficients

Gas	L–J Model		Buckingham Model			
	$\varepsilon/k(°K)$	$\sigma(\text{Å})$	$\varepsilon/k(°K)$	$r_m(\text{Å})$	α	$\dfrac{r_{max}}{r_m}$
Ne	34·9	2·78	38·0	3·147	14·5	0·185
A	119·8	3·405	123·2	3·866	14·0	0·203
Kr	171	3·60	158·3	4·056	12·3	0·28
CH$_4$	148·2	3·817	152·8	4·206	14·0	0·203

Symbols for Chapter 3

a_{22}, a_{24}, a_{14}	Coefficients	$P(r)$	Polynominal in r
b	Impact parameter	r_c	Relaxation distance
B	Positive constant	r_0	Distance of closest approach in atom–atom collision
C_1	Positive constant	s	Parameter
C_2	Positive constant	$U(R)$	Nuclear potential function
C	Positive constant	V	Electrostatic potential
E	Electric field	α	Parameter in Buckingham Potential
F	Force on molecule	δ	Positive constant
$f(\theta)$	Scattering amplitude at angle θ	δ_l	Phase shift
m	Reduced mass of atom	χ_l	Radial wave function

CHAPTER 4

The Pair Potential Function for Liquid Metals

4.1. Idealized Model for a Metal

A metallic system is distinguished from other systems by its ductility and by having high values of the electrical and thermal conductivities. These arise from the nearly free motion of some of the electrons—known as the conduction electrons. The remaining electrons (called the core electrons) are bound to the nucleus and the whole is termed an ion: the number, Z, of electronic changes on an ion is equal to the number of conduction electrons per atom, which is usually set equal to the chemical valence. Thus a liquid metal may be thought of as two intermingled fluids, one composed of the ions and another composed of the conduction electron gas. If these two "fluids" had to be treated separately and the predicted effects modified by mutual interactions, the theoretical difficulties in obtaining a self-consistent treatment would be huge. Moreover, since the electron gas only exists when many metal atoms are brought together to form a dense liquid, there must be many body forces present, and at first sight a pair potential discussion would seem inappropriate. However in recent years the concept of a neutral "pseudo-atom" has evolved, which reduces the problem of interatomic forces to one of (essentially) the same type as considered in previous Chapters. The conduction electrons distribute themselves around each ion to form screening clouds and a "pseudo-atom" is the ion plus its screening charge. This point of view does not imply that the conduction electrons are localized, rather a part of a general charge cloud is allocated locally to each ion. This theory is not so well founded as that used for the forces in non-conducting liquids, but it has the advantages of being based on a clear picture of a metal and of fitting into the conventional liquid theory based on pair interactions.

In discussing interatomic forces, the metal will be treated as a liquid system composed of neutral pseudo-atoms, but (in contrast to Chapter 3) it will be convenient to carry through most of the analysis in momentum space. A basic reason for this is that the conduction electrons form a highly degenerate Fermi gas, which may be described conveniently in momentum space.[3,4] It can be shown on this picture that the momenta of the conduction electrons (to a first approximation) fill a sphere in momentum space which has a radius k_f—

$$k_f = (3\pi^2 Z\rho)^{\frac{1}{3}} \qquad (4.1)$$

It is assumed here that the Fermi surface (i.e. the surface of the sphere) is sharp; that is that its width $\Delta k_f \ll k_f$. Note also that equation (4.1) shows that k_f is a function of ρ, i.e. of atomic volume. The electronic levels are filled from zero to an energy called the Fermi energy (E_f) and given by—

$$E_f = \frac{\hbar^2 k_f^2}{2m_e} = \frac{\hbar^2}{2m_e}(3\pi^2 Z\rho)^{\frac{2}{3}} \qquad (4.2)$$

A momentum space discussion is used because the contribution to the interatomic forces arising from the electron–electron interaction is conveniently expressed in terms of k_f. Other contributions to the interatomic forces come from ion-electron interactions and ion–ion (coulomb) interactions. It will be realized that these interactions have a different origin and so are of a different type to those discussed in Chapter 3. Their treatment follows that of classical theory, and the interaction between the ions themselves and between the ions and electrons will be assumed to be coulombic (with small quantum-theory corrections). In particular the field between the ions and the electrons is equivalent to the classical "displacement" D. As in the classical theory, this is modified by the presence of a dielectric and the actual field, E, is D/ε (ε is the dielectric constant). This modification is obtained at equation (4.19) as the result of the interaction between the electrons themselves. The ion–electron and electron–electron interactions will be considered separately before the whole of $u(r)$ is discussed. After this the modifications to the ideal model, which must occur in a real liquid metal, will be considered.

4.2. The Ion–Electron Interaction

This discussion will be presented in outline only and follows the treatment given by Harrison.[13] It starts from the Schrödinger wave equation for electrons moving in an "average" field in the metal. The field is a "self-consistent" one, that is it must correctly predict the electronic states upon which it depends itself. Let $V(\mathbf{r})$ be the self consistent (Hartree) potential seen by each electron, then the wave equation for the electrons is—

$$\left(-\frac{\hbar^2 \nabla^2}{2m_e} + V(\mathbf{r})\right)\psi_i = E_i\psi_i \tag{4.3}$$

where E_i is energy of the ith state. These states may be separated into (i) the core states and (ii) the conduction band states denoted by the wave vector $\mathbf{k}$. For the latter states equation (4.3) may be shown (Harrison,[13] Section 1.3) to be equivalent to the equation—

$$\left(-\frac{\hbar^2 \nabla^2}{2m_e} + W(\mathbf{r})\right)\phi_{\mathbf{k}} = E_{\mathbf{k}}\phi_{\mathbf{k}} \tag{4.4}$$

where—

$$W(\mathbf{r}) = V(\mathbf{r}) + \left[E_{\mathbf{k}} + \frac{\hbar^2 \nabla^2}{2m_e} - V(\mathbf{r})\right]P$$

and—

$$\psi_{\mathbf{k}} = (1-P)\phi_{\mathbf{k}}$$

The projection operator, P, projects any function onto the core states and thus produces cancellation between the two terms in $W(\mathbf{r})$. It can be seen that equation (4.4) has the usual form of a wave equation, so that the original problem has been translated into an equivalent one involving a "pseudo-potential" $W(\mathbf{r})$ and a "pseudo-wave function" $\phi(\mathbf{r})$. The advantage of this step is that the pseudo-potential is a weak potential [because the central part of $V(\mathbf{r})$ is cancelled by the second term in $W(\mathbf{r})$], and consequently $W(\mathbf{r})$ can be used in perturbation calculations. However there is an important difference in that $W(\mathbf{r})$ is an operator whereas $V(\mathbf{r})$ is not: the importance of this difference in the present case will be emphasized below. The definition of $W(\mathbf{r})$ is arbitrary [since it is only required to be weak compared to $V(\mathbf{r})$] and thus the pseudo-potential is not a uniquely defined physical quantity, but is rather a mathematically useful device having an approximate physical meaning.

By considering all the individual terms entering $W(\mathbf{r})$ and showing that each depends only upon the ion position ($\mathbf{r}_j$) through the difference $\mathbf{r} - \mathbf{r}_j$, it is shown that—

$$W(\mathbf{r}) = \sum_j w(\mathbf{r} - \mathbf{r}_j) \tag{4.5}$$

where $w(\mathbf{r})$ is an individual ionic pseudo-potential. As indicated above it is convenient to work in Fourier (or momentum) space, and thus the transform of $w(\mathbf{r})$ is required, i.e.—

$$w(\mathbf{k}, \mathbf{q}) = \rho \int e^{-i(\mathbf{k}+\mathbf{q})\cdot\mathbf{r}} w(\mathbf{r}) e^{i\mathbf{k}\cdot\mathbf{r}} \, d\tau \qquad (4.6)$$

where $d\tau$ is a volume element. Since $w(\mathbf{r})$ is an operator, the factors in the integral may not be commuted, and thus $w(\mathbf{k}, \mathbf{q})$ depends upon the wave vectors of both the initial state $(\mathbf{k})$ and the final state $(\mathbf{k}+\mathbf{q})$. In order to simplify the problem, it will be assumed that $w(\mathbf{k}, \mathbf{q})$ is independent of $\mathbf{k}$, that is that $w(\mathbf{r})$ is a simple potential and the factors in equation (4.6) commute. This assumption is necessary in order to preserve the physical picture of a pseudo pair potential, and it will be valid if over the (limited) region of $\mathbf{k}$ space important to any problem $w(\mathbf{k}, \mathbf{q})$ is nearly independent of $\mathbf{k}$. Alternatively some average value of w may be used, but in this case the actual values of $w(\mathbf{q})$ may vary from problem to problem and the simplicity of this approach partially lost.

A full quantum mechanical analysis is normally employed to calculate the pseudo-potential, but necessarily this is a rather involved procedure and leads to complicated functional forms. In order to bring out those physical features that are important, a number of models have been put forward and one only of these will be described here. In view of the theoretical difficulties described above and in Section 4.5, a deeper treatment does not seem to be worthwhile for the problems considered here. Harrison's[13] model for the function $w(\mathbf{q})$ consists of two terms. First there is the coulomb attraction between the electron and the ion, and if the charge on the ion is assumed to be a point charge of magnitude Ze, this term is—

$$-\frac{4\pi Z e^2}{q^2} \equiv -\int \frac{Ze^2}{r} e^{-i\mathbf{q}\cdot\mathbf{r}} \, d\tau \qquad (4.7)$$

Secondly, there is a repulsion arising because the conduction electrons are excluded from the core essentially because of the Pauli exclusion principle. This effect will have a range r_c ($\sim$ the Bohr radius) and is assumed to fall off exponentially. Thus the second term is—

$$\frac{\beta}{[1+(qr_c)^2]^2} \propto \int e^{-r/r_c} e^{-i\mathbf{q}\cdot\mathbf{r}} \, d\tau \qquad (4.8)$$

where β is a constant to be determined. Thus the assumed form for $w(\mathbf{q})$, considering these two effects only (indicated by the superscript 0) is—

$$w^0(\mathbf{q}) = -\frac{4\pi Z e^2}{k_f^2}\left[\frac{k_f^2}{q^2} - \frac{\beta}{[1+(qr_c)^2]^2}\right] \qquad (4.9)$$

This model reduces to a "point ion model" if $(qr_c)^2 \ll 1$; in the limit of large q, however, the $(qr_c)^2$ term is required to make the pseudo-potential fall to zero. The parameter β may be found by fitting equation (4.9) to the result of a detailed calculation of $w^0(q)$—see Table 4.1 for actual values—and in this way it has been shown that β is independent of volume to a first approximation [that is the variation for a 10% change of volume is significantly less than the variation of β obtained from different calculations of $w^0(\mathbf{q})$]. Other pseudo-potential models have been described that rely on experimental data. For example the model of Heine and Abarenkov[14] uses data on atomic energy levels to describe the ion core potential and wave functions, but detailed models of this kind are beyond the scope of this book.

In this discussion the interaction between the conduction electrons has not been taken into account, and equation (4.9) needs to be modified to include this effect. Fortunately this may be done without modifying the original part of this theory.

4.3. The Electron–Electron Interaction

Each electron interacts with the others through a coulomb potential, $\Phi(r)$, which may be determined from the charge density, $n(r)$, due to all the electrons. These quantities are connected through Poissons equation, i.e.—

$$\nabla^2 \Phi(r) = -4\pi e^2 n(r) \qquad (4.10)$$

On Fourier transforming, this equation gives the simple result—

$$q^2 \Phi(q) = 4\pi e^2 n(q) \qquad (4.11)$$

The electron density must be calculated from the electronic wave functions (ϕ_k) which are solutions of equation (4.4). Since $W(r)$ is a weak potential an expansion to first order is made in $2m_e W/\hbar^2 k^2$, and the wave functions are written as a sum of plane waves—

$$\phi_k = \rho^{\frac{1}{2}} \sum_q a_q(\mathbf{k}) \, e^{i(\mathbf{k}+\mathbf{q}).\mathbf{r}} \qquad (4.12)$$

where $a_0 = 1$. The expansion for the electron density is, to first order (Harrison,[13] equation 2.39)—

$$\phi^*\phi = \rho\{1 + \sum_q'[a_q(\mathbf{k}) \, e^{i\mathbf{q}.\mathbf{r}} + a_q^*(\mathbf{k}) \, e^{-i\mathbf{q}.\mathbf{r}}] + \text{etc.}\} \qquad (4.13)$$

where the prime indicates that the $q = 0$ term is omitted. The first term in this expansion gives the uniform negative charge which cancels the positive charge of the ions: the second term gives the fluctuations from the average which modify the simple ion–electron potential. The fluctuating term may be divided into components associated with each ion, and hence each component produces a screening field. It may be shown (Harrison,[13] equation 2.10) from the expansion of equation (4.4) that to the same order of approximation, $a_q(\mathbf{k})$ is related to the pseudo-potential $W(\mathbf{k}, \mathbf{q})$ by—

$$a_q(\mathbf{k})\frac{\hbar^2}{2m_e}(k^2 - |\mathbf{k}+\mathbf{q}|^2) \simeq W(\mathbf{k}, \mathbf{q}) \qquad (4.14)$$

(here W covers the effects included in equation (4.9) and also the electron–electron interaction).

Now the electron density required in equation (4.11) is given by the Fourier transform of the fluctuating part of $\phi^*\phi$, and hence is obtained in terms of $a_q(\mathbf{k})$ as (an integration over the Fermi sphere is denoted by $\int_F$)—

$$n(\mathbf{q}) = \frac{4}{(2\pi)^3}\int_F d\mathbf{k}\, a_q(\mathbf{k}) \qquad (4.15)$$

At this stage it is necessary to assume (as in Section 4.2) that $W(\mathbf{k}, \mathbf{q})$ is independent of $\mathbf{k}$, in other words $W(r)$ is a simple or local potential. Thus combining equations (4.14), and (4.15) with this assumption gives—

$$n(q) = \frac{8m_e W(q)}{(2\pi)^3\hbar^2}\int_F \frac{d\mathbf{k}}{k^2 - |\mathbf{k}+\mathbf{q}|^2} = \frac{q^2 W(q)}{4\pi e^2}[1 - \varepsilon(q)] \qquad (4.16)$$

where—

$$\varepsilon(q) = 1 + \frac{2m_e e^2 k_f}{\pi\hbar^2 q^2}\left(1 + \frac{4k_f^2 - q^2}{4k_f q}\log_e\left|\frac{2k_f - q}{2k_f + q}\right|\right) \qquad (4.17)$$

$\varepsilon(q)$ is called the "dielectric function", and the potential $W(q)$ may be related to $W^0(q)$ through this function. First equations (4.16) and (4.11) are combined—

$$\Phi(q) = W(q)[1 - \varepsilon(q)] \qquad (4.18)$$

and then $W(q)$ is written as a sum of the unscreened electron–ion interaction, $W^0(q)$, and the electron–electron interaction, i.e.—

$$W(q) = W^0(q) + \Phi(q) = W^0(q)/\varepsilon(q) \qquad (4.19)$$

Thus the electron–electron interaction can be included in a simple way through the use of the dielectric function, the form of which has been chosen so that equation (4.19) is the analogue of the classical equation $D = \varepsilon E$.

4.4. The Effective Interatomic Potential

For the purposes of calculation the potential energy $u_p(r)$ between two pseudo-atoms is divided into two parts—

$$u_p(r) = u_d(r) + u_i(r) \tag{4.20}$$

where u_d is a direct interaction between the ions, and u_i is an "indirect" interaction which arises when the ions are placed in a bath of electrons (i.e. due to the terms calculated in Sections 4.2 and 4.3). Now if the ions do not overlap the direct interaction can be approximated by the Coulomb interaction between point ions of charge Z^*e, where Z^* is an effective valence given by[13]—

$$Z^* = Z + Z_1 \tag{4.21}$$

The second term arises from that part of the conduction electron density which is localized at each ion, and in most cases $Z_1 < 0.1Z$. Thus the direct term is—

$$u_d(q) = \frac{4\pi e^2 Z^{*2}}{q^2} \tag{4.22}$$

Possibly a term due to overlap repulsion should be added to equation (4.22) but the electrostatic repulsion is expected to be the dominant term over the range of r of interest here (see range of Fig. 4.1). In order to check this term ion–atom or ion–ion scattering experiments[12,16] may be carried out, and such experiments have confirmed the form of equation (4.22).

The term $u_i(q)$ is the modification of the direct term caused by placing the ions in a bath of electrons, and so it is the difference between the interatomic energy calculated by using equation (4.19) for the pseudo-potential and the same thing using $\varepsilon = 1$. In order to find these energies the charge density $Z^0(q)$ equivalent to the un-screened pseudo-potential $w^0(q)$ is calculated from Poissons equation (4.11), i.e.—

$$Z^0(q) = \frac{q^2}{4\pi e^2} w^0(q) \tag{4.23}$$

Now transforming to real space the field produced by the charge density $Z^0(\mathbf{r}')$ at a point $\mathbf{r} - \mathbf{r}'$ is the screened pseudo-potential $w(\mathbf{r} - \mathbf{r}')$. Consequently the interaction between two pseudo-ions at $\mathbf{r}'$ and $(\mathbf{r} - \mathbf{r}')$ is—

$$u_0(r) = \int w(\mathbf{r} - \mathbf{r}') Z^0(\mathbf{r}') \, d\mathbf{r}' \tag{4.24}$$

From this simple point of view there should be two terms in equation (4.24)—as each ion interacts with the field produced by the other—

but[13] in the self-consistent-field treatment the sum of the energies of each of the electrons counts their interaction energy twice so that the second term is not required. Equation (4.24) is readily transformed to give—

$$u_0(q) = w(q)Z^0(q) = \frac{[w^0(q)]^2 q^2}{4\pi e^2 \varepsilon(q)}$$ (4.25)

by using equations (4.23) and (4.19). Thus finally the indirect term is—

$$u_i(q) = \frac{[w^0(q)]^2 q^2}{4\pi e^2}\left(\frac{1}{\varepsilon(q)} - 1\right)$$ (4.26)

and the potential energy between two pseudo-atoms is [from equations (4.20), (4.22), (4.26) and (4.9)]—

$$u_p(r) = 4\pi e^2 Z^2 \int e^{i\mathbf{q}\cdot\mathbf{r}} \, d\mathbf{q}\left[\frac{Z^{*2}}{q^2 Z^2} + \frac{q^2}{k_f^4}\left(\frac{k_f^2}{q^2} - \frac{\beta}{[1+(qr_c)^2]^2}\right)^2\left(\frac{1}{\varepsilon(q)} - 1\right)\right]$$ (4.27)

Equation (4.27) is based on a variety of approximations and assumptions, the cumulative effect of which is uncertain. But (assuming r_c is approximately 0.5 Å $\sim \hbar^2/m_e e^2$) it depends upon only one unknown, β, and the value of this parameter can be estimated theoretically (Section 4.6).

Equation (4.27) is illustrated in Fig. 4.1, which shows the calculated potential for aluminium. At large r this equation reduces to an

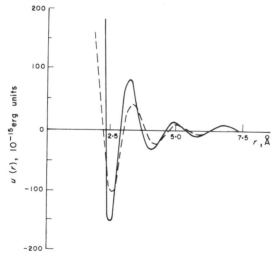

FIG. 4.1. Effective inter-atomic potential in aluminium according to equation (4.27). The full line is the result of a complete calculation, and the dotted line is equation (4.28) (from Harrison,[13] Fig. 2.3).

oscillatory form given by—

$$u_p(r) \simeq \frac{9\pi Z^2 [w^0(2k_f)]^2}{E_f} \cdot \frac{\cos 2k_f r}{(2k_f r)^3} \tag{4.28}$$

which is plotted in Fig. 4.1 and is a rough approximation to the complete result. The oscillatory term arises from the Fourier transformation of the logarithmic term in $\varepsilon(q)$—equation (4.17)—and is directly related to the assumption that the Fermi surface is sharply defined.

4.5. Pair Potential in a Real Metal

The theory discussed above was based on an idealized model of a liquid metal, and consequently the resulting pair potential must be regarded as an idealized result. Perhaps its most important defect may be seen by comparing its large r behaviour (equation 4.28) with the basic requirement of equation (3.2). Such a comparison shows that the oscillations given by equation (4.28) should be damped, and this behaviour might be expected in practice, since the Fermi surface cannot be perfectly sharp. From this point of view, the Coulomb forces are not long range due to self cancellation. It is usual to assume that the damping is exponential at large r so that the actual potential might have the form $e^{-\alpha r} u_p(r)$, where α is an unknown constant. An expression of this type will damp faster than an inverse-power law, and consequently at large r a weak interaction, giving a power law contribution, might be important. There are several points of view on this question hinging upon the extent to which the conduction electrons can screen charges and dipoles simultaneously. If simultaneous screening occurs the exponential damping factor will apply to both types of interaction and the long range behaviour will have the form $e^{-\alpha r} u_p(r)$. However if, at any instant, the charges are screened but the instantaneous dipoles are not, a term may occur in the potential which varies as an inverse power of r. Possible interactions include the polarization attraction between a neutral atom and a charge (r^{-4}), the attractive force between charges and dipoles (r^{-5}) and the dipolar force (r^{-6}). Thus if the pair potential approximation is valid in a metal the actual potential might have the form—

$$u(r) \simeq e^{-\alpha r} u_p(r) - \frac{C}{r^s} \tag{4.29}$$

where $s \geq 4$ and C is a small constant. An illustration of the possible form for equation (4.29) is given in Fig. 4.2; the shape of u_p is retained at low values of r, but at large r this term is damped and weak long range forces become apparent.

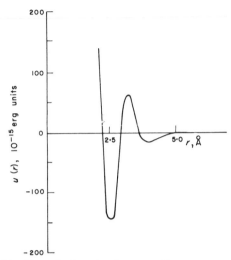

FIG. 4.2. Probable effect of a damping factor and long-range term in $u(r)$ is to modify potential of Fig. 4.1 to that shown here.

Other difficulties arise in the use of $u_p(r)$ for temperature or volume dependent problems. Since k_f (equation 4.1) is volume dependent it follows that $u_p(r)$ is weakly volume dependent and so does not satisfy the requirements of an ideal pair potential. As the volume is increased the electrical resistivity increases and in the limit of low volume the metallic properties disappear. This effect may be included in equation (4.29) by allowing α and C to increase (and perhaps s and Z to vary) as the volume increases, and eventually a term due to overlap repulsion should be added when the Coulomb repulsion becomes small. In this limit the potential for the metal has been transformed to the form for a non-conducting liquid. The above description is essentially a description of many body forces in a liquid metal in terms of the behaviour of a limited number of parameters that appear in an arbitrary pair potential.

4.6. Evaluation of the Repulsive Parameter β

Even after the modifications of Section 4.5, the idealized potential (equation 4.27) gives an important contribution to $u(r)$ as shown in Fig. 4.2. It is, therefore, important to derive the values of the parameters (β and r_c) appearing in $u_p(r)$. Of these two parameters r_c is known approximately—Section 4.4—but β is unknown, and this section is devoted to methods of obtaining values of β.

Examination of equation (4.9) shows that at both high and low q, $w^0(q)$ is insensitive to the value of β. Thus a convenient point at which to find the value of β is the first zero of $w^0(q)$. If $q = q_0$ here then—

$$\beta = \frac{[1+(q_0 r_c)^2]^2 k_f^2}{q_0^2} \tag{4.30}$$

The value of q_0 may be found from a complete theoretical calculation[13] of $w^0(q)$: as a first approximation it may be set equal to $1 \cdot 5\, k_f$, since it is found to be near this value in many cases. Thus the approximation $q_0 = 1.5\, k_f$ in equation (4.30) provides the first approximate method of finding β.

A method of finding β experimentally from the electrical resistivity has the advantage of providing a value which is consistent with the spirit of the theory. In measuring the resistivity the conduction electrons are moved through the metal and the resistance observed is related to the pseudo-potential and hence the value of β. Since $w(r)$ is a weak potential the Born approximation formula (equation 3.30) is normally used for the electron scattering cross-section. Moreover since[15] only those electrons for which $k = k_f$ contribute to the resistivity, the value of $w(\mathbf{q}, \mathbf{k})$ at equation (4.6) need be evaluated at $k = k_f$ only. If w is really independent of k this value will apply to all k and hence give the correct value of β in the present case. Thus this method includes the error in the assumption that $w(r)$ is a local potential and the error in calculating the electron scattering cross-section from the Born approximation (or an alternative approximation if preferred). It is assumed now that the electrons are scattered with negligible energy loss and thus the scattering angle, θ, is related to the wave vector change q by—

$$q = 2k_f \sin \theta/2 \tag{4.31}$$

and the limits of q are $(0, 2k_f)$. Thus the scattering cross-section for a system of ions is obtained from equations (3.49) and (3.23) by averaging the amplitude $f(\theta)$ over all the ions, and squaring the result—

$$\sigma(\theta) = \left| \frac{1}{N} \sum_i \frac{1}{4\pi} \int e^{i\mathbf{q} \cdot \mathbf{r}_i} w(r_i)\, d\mathbf{r}_i \right|^2 = S(q)|w(q)|^2 \tag{4.32}$$

where—

$$S(q) = \frac{1}{N} \sum e^{i\mathbf{q} \cdot (\mathbf{r}_i - \mathbf{r}_j)}$$

is related to the Fourier transform of $g(r)$ (equation 6.22). Finally the resistivity (ρ_e) is given[15] from normal transport theory as—

$$\rho_e = \frac{2\pi m_e}{ZNe^2} \int_0^\pi \sigma(\theta)(1 - \cos\theta) \sin\theta \, d\theta \qquad (4.33)$$

If the structure factor $S(q)$ is known (see Chapter 6) then ρ_e may be calculated from equation (4.33) and compared to experiment, or conversely equation (4.33) may be used to derive a value of β from ρ_e.

TABLE 4.1
Model parameters for liquid metals

Metal	$k_f(\text{Å})$	β from equation (4.30)*	β from fitting to calculated pseudo-potential	Tempera-ture for ρ_e (°C)	Calculated ρ_e	Measured ρ_e
					ohm-cm	
(1)	(2)	(3)	(4)	(5)	(6)	(7)
Na	0·89	0·50	0·53	100	9·4	9·6
Al	1·75	0·65	0·77	700	24·5	24·7
Sn	1·61	0·62	0·60	—	—	—
Pb	1·57	0·62	0·68	350	58	96

* By using the values $r_c = 0\cdot33\,\text{Å}$; $q_o = 1\cdot5\,k_f$.

Table 4.1† summarizes the values of β derived from equation (4.30)—column 3—and by fitting to calculations of $w(q)$—column 4. Also shown are the values of the electrical resistivities calculated from equation (4.33) by using the calculated values of $w(q)$—column 6. The fair agreement between the theoretical and experimental values for ρ_e give some justification for the choice of β in column 4.

Finally it should be pointed out that the electronic component (κ_e) of the thermal conductivity is related[15] to the electrical resistivity by the approximate relation—

$$\kappa_e \simeq \frac{1}{\rho_e} \cdot \frac{\pi^2 T}{3} \left(\frac{k}{e}\right)^2 \qquad (4.34)$$

where k is Boltzmanns constant. This is, of course, just a statement of the Weideman–Franz–Lorenz law: that it is not exact can be seen

† Table 4.1 is compiled from data given in ref. 13.

from the fact that it does not account for the variation of $\kappa_e \rho_e$ across the melting point (for example). Thus, in principle, the thermal conductivity could be used (with equations 4.34 and 4.33) in place of the resistivity to evaluate β, since more than 99 % of it is contributed by the electronic effects in a typical metal.

Symbols for Chapter 4

a_q	Amplitude of q component of electronic wave function	$W(r)$	Pseudo-potential seen by electron in a metal [$w°(r)$ for one ion]
E_k	Energy of electron of wave number k	Z^*	Effective valence
$n(r)$	Charge density	$z°$	Charge density corresponding to w^0
P	Projection operator	κ_e	Thermal conductivity due to electrons
u_i, u_d	Indirect and direct parts of $u(r)$		
$u_p(r)$	Interaction between two pseudo-ions	ρ_e	Electrical resistivity
$V(r)$	Self consistent potential seen by electrons in a metal	$\Phi(r)$	Coulomb potential
		ϕ	Pseudo electronic wave function
$W°(r)$	Unscreened potential seen by electron in a metal [$w(r)$ for one ion]	ψ	Electronic wave function

CHAPTER 5

Relations between $g(r)$ and $u(r)$

It was shown in Chapters 3 and 4 that the pair potential, $u(r)$, can be calculated, in principle, from the basic properties of the atoms composing the liquid. For this reason $u(r)$ is regarded as a basic quantity from which all other liquid properties should be calculated. In this Chapter the methods of calculating $g(r)$ (equation 2.9) from $u(r)$ are considered, and if these methods are successful it would be possible to calculate the equation of state (for example) from first principles. The first step is to derive a relation between $n^{(2)}$, $n^{(3)}$ and $u(r)$ in a classical system. Then if a suitable approximation is made for $n^{(3)}$ this relation will reduce to one between $u(r)$ and $g(r)$ only. Several approximations have been suggested for this purpose, and one of them will be described (Section 5.2). In other cases, the approximation is expressed more conveniently through a different function (the so called "direct" correlation function) and two examples of this procedure will be given (Section 5.3 to 5.6). Finally (Section 5.8) a method of calculating $g(r)$ from $u(r)$ employing a model for the liquid will be described. Although it was suggested above that these relationships could be used to calculate $g(r)$ from $u(r)$, they may be used also to obtain $u(r)$ from $g(r)$ if $g(r)$ has been measured experimentally (experimental methods are described in Chapter 6). The discussion of these two applications of the $g(r)$–$u(r)$ relations will be given in Chapter 7.

5.1. A General Relationship

Figure 5.1 shows the positions of any three atoms in the liquid and defines the vectors used in the discussion below. The mean force acting on molecule 1 at position $\mathbf{r}_1$ can be expressed in terms of $g(r)$ from equation (2.10b), i.e.—

$$\text{Mean force} = -\frac{\partial \phi(r)}{\partial \mathbf{r}_1} = \frac{kT}{g(r)} \frac{\partial g(r)}{\partial \mathbf{r}_1} \tag{5.1}$$

However the mean force may be evaluated in another way, as follows.

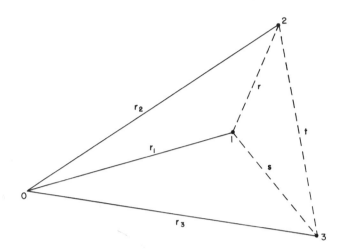

FIG. 5.1. Definition of vectors connecting any three atoms in the liquid. The origin is denoted by O, and the full circles mark the three atoms (indexed 1, 2 and 3).

The probability of a third molecule being in $d\mathbf{r}_3$ at $\mathbf{r}_3$ when the first two molecules are at $\mathbf{r}_1$ and $\mathbf{r}_2$ is—

$$p(\mathbf{r}_1, \mathbf{r}_2, \mathbf{r}_3)\, d\mathbf{r}_3 = \frac{n^{(3)}(\mathbf{r}_1, \mathbf{r}_2, \mathbf{r}_3)\, d\mathbf{r}_3}{n^{(2)}(\mathbf{r}_1, \mathbf{r}_2)} \qquad (5.2)$$

The force on molecule 1 due to molecule 3 is given by (where $\mathbf{s} = \mathbf{r}_3 - \mathbf{r}_1$)—

$$-\frac{\partial u(s)}{\partial \mathbf{r}_1} p(\mathbf{r}_1, \mathbf{r}_2, \mathbf{r}_3)\, d\mathbf{r}_3 \qquad (5.3)$$

and the force on molecule 1 due to molecule 2 is just $-[\partial u(r)/\partial \mathbf{r}_1]$. Thus the force on molecule 1 due to all other molecules is—

$$\text{Mean force} = -\frac{\partial u(r)}{\partial \mathbf{r}_1} - \int \frac{\partial u(s)}{\partial \mathbf{r}_1} p(\mathbf{r}_1, \mathbf{r}_2, \mathbf{r}_3)\, d\mathbf{r}_3 \qquad (5.4)$$

Combination of equations (5.1) and (5.4) gives the required general relationship, i.e.—

$$\frac{\partial g(r)}{\partial \mathbf{r}_1} \cdot \frac{kT}{g(r)} + \frac{\partial u(r)}{\partial \mathbf{r}_1} = -\int \frac{\partial u(s)}{\partial \mathbf{r}_1} p(\mathbf{r}_1, \mathbf{r}_2, \mathbf{r}_3)\, d\mathbf{r}_3 \qquad (5.5)$$

This derivation, via a physical picture, may be checked by substituting the definition of $n^{(2)}$ and $n^{(3)}$—equations (2.2) or (2.6a)—into equation (5.5). In addition the general definition, equation (2.2) or (2.6a), can be used to establish a recurrence relation between $n^{(r)}$ and $n^{(r+1)}$ so giving a hierachy of equations. However these equations cannot be solved—to give $n^{(2)}$ for example—without some other information. There have been a number of different suggestions on the best method to use and three of these will be discussed below. This discussion starts conveniently from equation (5.5) rewritten in the form—

$$\frac{\partial}{\partial r}\left\{\ln g(r)+\frac{u(r)}{kT}\right\} = -\frac{1}{kT}\int \frac{\partial u(s)}{\partial s}\frac{(\mathbf{r}.\mathbf{s})}{rs}.p(\mathbf{r}_1,\mathbf{r}_2,\mathbf{r}_3)\,d\mathbf{r}_3 \quad (5.6a)$$

or—

$$g(r) = \exp -\frac{u(r)+W(r)}{kT}$$

where—

$$W(r) = \int_r^\infty dr'\int_V \frac{\partial u(s)}{\partial s}\frac{\mathbf{r}'.\mathbf{s}}{r's}p(\mathbf{r}_1,\mathbf{r}_2,\mathbf{r}_3)\,d\mathbf{r}_3$$

$$\left.\right\} \quad (5.6b)$$

and $\mathbf{r}_1-\mathbf{r}_2$ is redefined as $\mathbf{r}'$.

If an approximation is made for $p(\mathbf{r}_1,\mathbf{r}_2,\mathbf{r}_3)$ then $W(r)$ and so $g(r)$ may be calculated. However it may be expected that $W(r)$ will depend upon $g(r)$—e.g. the Yvon–Born–Green method—so that an iterative method would have to be adopted to solve this equation. Two other methods, namely the hypernetted chain and Percus–Yevick approximations, do not involve explicit approximations to $p(\mathbf{r}_1,\mathbf{r}_2,\mathbf{r}_3)$. However as equation (5.6) is valid generally any approximate treatment of $g(r)$ implies an approximation for $W(r)$, and the accuracy of this approximation determines the validity of the treatment of $g(r)$.

5.2. The Yvon–Born–Green Equation (YBG)

An approximation for $n^{(3)}$, known as the superposition approximation and due to Kirkwood, was employed by Yvon and by Born and Green (see Rice and Gray[5]) to reduce equation (5.6). This states that the probability of finding molecule 3 at $\mathbf{r}_3$ molecule 2 at $\mathbf{r}_2$ and molecule 1 at $\mathbf{r}_1$ is the product of the probabilities of finding the three separate pairs of molecules (1, 2) and (2, 3) and (3, 1) at the appropriate spacings. If one molecule is so far away from the other two that its potential does not affect the total interaction of the three, then this will be a good approximation. Consequently it may be expected to be valid at low densities and become increasingly poor at high densities.

Specifically the superposition approximation is—

$$n^{(3)}(\mathbf{r}_1, \mathbf{r}_2, \mathbf{r}_3) \simeq \frac{n^{(2)}(r)n^{(2)}(s)n^{(2)}(t)}{\rho^3} \tag{5.7}$$

Since $g(r)$ is a function of density this approximation is not just the $\rho \to 0$ limit of $n^{(3)}$, but includes some terms to all orders in the density. After some reduction equation (5.6) with equation (5.7) becomes[17]—

$$\ln g(r) + \frac{u(r)}{kT} = -\rho \int_V E(|\mathbf{r}_1 - \mathbf{r}|)h(\mathbf{r}_1)\,\mathrm{d}\mathbf{r}_1 = -\frac{W(r)}{kT} \tag{5.8}$$

where—

$$E(r) = \frac{1}{kT} \int_\infty^r g(x)\frac{\partial u(x)}{\partial x}\,\mathrm{d}x \quad \text{and} \quad h(r) = g(r) - 1$$

Equation (5.8) is known as the YBG equation between $g(r)$ and $u(r)$.

5.3. The Hypernetted Chain Equation (HNC)

The HNC approximation is obtained by carrying out a cluster expansion for $n^{(3)}$ and then neglecting certain terms which make the summation of the expansion difficult. It is found[5] that the parts that are retained can be written conveniently in terms of another function, $c(r)$, called the "direct correlation function". The definition of $c(r)$ in terms of $h(r)$ is—

$$h(r) = c(r) + \rho \int c(|\mathbf{r} - \mathbf{r}'|)h(r')\,\mathrm{d}\mathbf{r}' \tag{5.9}$$

Equation (5.9) was originally introduced by Ornstein and Zernike, who argued that the "total correlation"—$h(r)$—was the sum of a "direct effect", $c(r)$, of molecule 1 on molecule 2 plus an indirect effect of all other molecules. The latter effect is expressed through the convolution of $h(r)$ and $c(r)$. A virtue of the function $c(r)$ is that in many circumstances it is short ranged in comparison with $h(r)$. The equation finally obtained by this method is[5]—

$$\ln g(r) + \frac{u(r)}{kT} = \rho \int c(|\mathbf{r} - \mathbf{r}'|)h(r')\,\mathrm{d}\mathbf{r}' = h(r) - c(r) = -\frac{W(r)}{kT} \tag{5.10}$$

This equation connecting $g(r)$ and $u(r)$, is accurate at low densities because only the higher-order terms in the cluster expansion are affected by the approximation. However, since the expansion is summed to all orders in the density, it can be used at high density and will give different results from the YBG equation.

5.4. The Percus Yevick Equation (PY)

The Percus Yevick equation may be obtained[5] in a similar manner to the HNC equation through a cluster expansion. Likewise it may be written conveniently in terms of the direct correlation function, i.e.—

$$\ln g(r) + \frac{u(r)}{kT} = \ln\{g(r) - c(r)\} = -\frac{W(r)}{kT} \qquad (5.11)$$

For the same reason as given above, this equation is accurate at low densities only, but as the expansion is summed to all orders in the density it may be useful at high density also. In both the HNC and PY approximations, it is hoped that the neglected terms are small or have a strong mutual cancellation, and hence are small in total effect.

5.5. Cluster Expansion for the Direct Correlation Function

It is convenient to discuss the HNC and PY approximations by examining the cluster expansion for the direct correlation function. This expansion can be shown[5] to be—

$$c(r) = [\text{O--O}] + \rho \left[\triangle \right]$$

$$+ \frac{\rho^2}{2} \left[2 \square + 4 \square + \square + \square \right] + \text{etc.} \qquad (5.12a)$$

where the diagrams have the following meaning—

$${}_1\text{O--O}_2 = f_{12} = \exp\left[-\frac{u(r_{12})}{kT}\right] - 1$$

$$\triangle = \int_V f_{12} f_{31} f_{23} \, dr_3$$

O indicates a subscript in f_{ij} and ● indicates a subscript and an integration.

and so on.

In the HNC and PY approximations, the first two terms of this expansion are retained, but the third and higher terms are truncated.

For example, the third term is—

$$\text{HNC} \quad \frac{\rho^2}{2}\left[2\;\square + 4\;\boxtimes + \boxtimes\right] \tag{5.12b}$$

$$\text{PY} \quad \frac{\rho^2}{2}\left[2\;\square + 4\;\boxtimes\right]$$

By comparison of equations (5.12a) and (5.12b) it can be seen that one diagram is dropped in the HNC case and two are dropped in the PY case. In spite of this the PY approximation is found to be superior when repulsive forces are dominant because the two diagrams dropped tend to cancel each other. Similar remarks apply to the higher-order terms in equation (5.12a). The effect of dropping these diagrams is to enable the series (equation 5.12a) to be summed, giving—

$$\text{HNC} \quad c(r) \simeq h(r) - \ln g(r) - u(r)/kT$$
$$\text{PY} \quad c(r) \simeq g(r)(1 - e^{u(r)/kT}) \tag{5.13a}$$

which are restatements of equations (5.10) and (5.11).

Useful expansion formulae are obtained by expanding the logarithm in the HNC equation and the exponential in the PY equation, i.e.—

$$\text{HNC} \quad c(r) = -\frac{u(r)}{kT} - \frac{h^2(r)}{2} + \text{etc.}$$
$$\text{PY} \quad c(r) = -\frac{u(r)}{kT} - h(r)\frac{u(r)}{kT} + \text{etc.} \tag{5.13b}$$

In both cases the leading term is the same and provided $h(r)$ is sufficiently small these series can be truncated at this term. For example if $r \to \infty$ this may be done. However, expansion of the YBG equation will not give the same limit. A useful expansion[17] for the YBG equation is to write—

$$g(r) = \exp\left(g_1(r) - \frac{u(r)}{kT}\right) \tag{5.14}$$

and expand in powers of $g_1(r)$. The linear term in this expansion is the same as obtained by expanding the HNC or PY equations in this way. These expressions are, effectively, low-density expansions, and their agreement confirms that the three starting equations are equivalent at low density.

5.6. Solution of the PY Equation for Hard Spheres

For those problems where the core of the potential is the important feature, it may be appropriate to approximate the actual potential by equation (3.12). Because of the relative accuracy of the PY solution (Fig. 7.3) it is helpful to obtain an exact solution of it for this case. Wertheim[18] has given a solution as follows. Write the radial distribution function as—

$$g(r) = \exp\left(-\frac{u(r)}{kT}\right) \cdot \tau(r) \tag{5.15}$$

so that for the PY assumption (equation 5.13)—

$$c(r) = \tau(r)\left(\exp\left(-\frac{u(r)}{kT}\right) - 1\right)$$

Then the equation defining $c(r)$ (equation 5.9) becomes—

$$h(r) - c(r) = \tau(r) - 1 = \int h(|\mathbf{r} - \mathbf{r}'|)c(r')\,d\mathbf{r}'$$

$$= -\rho\int\tau(r')[e^{-u(r)/kT} - 1]\,d\mathbf{r}' + \rho\int e^{-u(\mathbf{r}-\mathbf{r}')/kT}\tau(|\mathbf{r} - \mathbf{r}'|)\tau(r')$$

$$\times [e^{-u(r')/kT} - 1]\,d\mathbf{r}' \tag{5.16}$$

Upon inserting the hard-sphere potential (equation 3.12), this result becomes—

$$\tau(r) = 1 + \rho\int_{r'<\sigma}\tau(r')\,d\mathbf{r}' - \rho\int_{\substack{r'<\sigma \\ |\mathbf{r}-\mathbf{r}'|>\sigma}}\tau(r')\tau(|\mathbf{r}-\mathbf{r}'|)\,d\mathbf{r}' \tag{5.17}$$

where σ is the radius of the sphere. Thus the problem is to solve (5.17) for $\tau(r)$ and then show that the $\tau(r)$ so obtained is the only solution having the correct physical properties. Only the former question will be discussed here. Equation (5.17) may be Laplace transformed to—

$$t[F(t) + G(t)] = \frac{1}{t}\left[1 + \frac{24\eta}{\sigma^3}\int_0^\sigma \tau(r)r^2\,dr\right] - 12\eta[F(-t) - F(t)]G(t) \tag{5.18}$$

where—

$$F(t) = \frac{1}{\sigma^2}\int_0^\sigma r\tau(r)\,e^{-tr/\sigma}\,dr$$

$$G(t) = \frac{1}{\sigma^2}\int_\sigma^\infty r\tau(r)\,e^{-tr/\sigma}\,dr$$
$$\left.\right\} \tag{5.19}$$

and—

$$\eta = \tfrac{1}{6}\pi\sigma^3\rho$$

A solution of the form—

$$\tau(r) = -c(r) = \alpha + \frac{\beta r}{\sigma} + \gamma \left(\frac{r}{\sigma}\right)^2 + \delta \left(\frac{r}{\sigma}\right)^3 \tag{5.20}$$

for $r < \sigma$, is assumed to hold (the reasons are discussed by Wertheim[18]). This trial solution enables $F(t)$ to be calculated (equation 5.19) in terms of α, β, γ and δ, and then from equation (5.18) $G(t)$ is obtained in terms of the same quantities. Finally it can be shown[18] from equation (5.17) that $\tau^{(n)}(r)$—n denotes differentiation n times—is continuous at $r = \sigma$ for $n = 0$, 1, 2, and $G(t)$ is expanded in a series in $\tau^{(n)}(\sigma)/t^{n+1}$ for large values of t. The coefficients in this series are equated with the coefficients in the corresponding $1/t$ expansion of the known function $G(t)$, thus giving three algebraic equations for α, β, γ and δ. A fourth equation is obtained by putting $r = 0$ in equation (5.17) to obtain—

$$\tau(0) = 1 + \frac{24\eta}{\sigma^3} \int_0^\sigma \tau(r) r^2 \, dr \tag{5.21}$$

Solution of these four equations then gives the values of α, β, γ and δ ($\gamma = 0$, but the others are finite) and hence the complete expression for $c(r)$, i.e.—

$$c_{\text{ph}}(r) = -\frac{1}{(1-\eta)^4} \left[(1+2\eta)^2 - 6\eta \left(1+\frac{\eta}{2}\right)^2 \frac{r}{\sigma} + \frac{\eta}{2}(1+2\eta)^2 \left(\frac{r}{\sigma}\right)^3 \right] \quad \text{for } r < \sigma \tag{5.22}$$

and—

$$c(r) = 0 \text{ for } r \geq \sigma$$

where the subscript (ph) denotes "Percus–Yevick hard sphere" solution. All other relevant functions may be obtained from $c(r)$, but discussion of them will be deferred until later.

5.7. Physical Significance of Equation (5.6)

In order to understand the effect of the various terms in equation (5.6), it is useful to employ a simple model for $p(\mathbf{r}_1, \mathbf{r}_2, \mathbf{r}_3)$. The model used here may be written—

$$p(\mathbf{r}_1, \mathbf{r}_2, \mathbf{r}_3) = \rho g'(s) g'(t) \tag{5.23a}$$

where—

$$\left. \begin{array}{ll} g'(x) = 0 & x \leq \sigma_1 \\ g'(x) = 1 & x > \sigma_1 \end{array} \right\} \tag{5.23b}$$

Figure 5.2 shows the physical situation depicted by equation (5.23); two spherical "sockets" of radius σ_1 are placed around atoms 1 and 2, and for small r they are allowed to overlap. In the course of the integration of third-atom positions over the volume of the liquid the third atom is excluded from these sockets (but it has a uniform probability of being elsewhere). From the symmetry of the integration this is equivalent to allowing the third atom to occupy uniformly the dotted

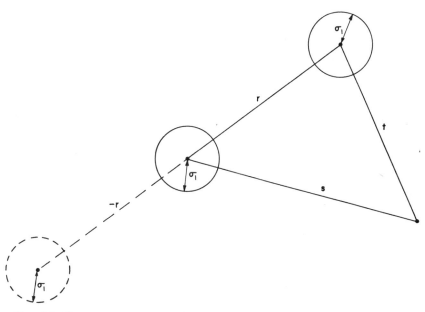

FIG. 5.2. Illustration of the effect of excluded volume or the "third atom" force. Atom 3 is allowed to move over the whole volume except the region (shown by the circles) occupied by atoms 1 and 2. By symmetry this is equivalent to allowing atom 3 to occupy the dotted circle only, which produces a repulsive force between 1 and 2 overcome by the attraction of 1 and 2 only when they are close together.

socket at $-\mathbf{r}$ but be excluded from all other regions. This is the situation depicted by equation (5.8), where from equation (5.23) and the definition of $E(x)$—

$$\begin{aligned}E(x) &= u(x)/kT \qquad r \geq \sigma_1 \\ &= u(\sigma_1)/kT \qquad r < \sigma_1\end{aligned}\Bigg\} \qquad (5.24)$$

Clearly a model of this kind emphasizes the core of the potential and so can only be valid if the third atom is close to both of the other two: that is for all three atoms to be close together. For the present purpose of illustration, this will be taken as r small.

If equation (5.6) is simplified by equation (5.23), it may be integrated to give—

$$g(r) = \exp -\frac{u(r) + W_s(r)}{kT} \tag{5.25a}$$

where—

$$W_s(r) = -\frac{\pi \rho}{r} \int_l^{r+a} \frac{\partial u(s)}{\partial s} \left[\frac{\sigma_1^4}{4} - \frac{2r\sigma_1^3}{3} + \frac{\sigma_1^2 r}{2} - \frac{r^4}{12} + \frac{s^2(r^2 - \sigma_1^2)}{2} \right.$$
$$\left. - \frac{2s^3 r}{3} + \frac{s^4}{4} \right] ds \tag{5.25b}$$

and—

$$l = r - \sigma_1 \quad \text{for } r \geq 2\sigma_1$$
$$= \sigma_1 \quad \text{for } r < 2\sigma_1$$

As an illustration the function $W_s(r)$ has been evaluated from equation (5.25b) for the Lennard–Jones potential by using the parameters for argon (Table 3.1) and the density of argon at a temperature of 84·5°K. Figure 5.3 shows the result of this calculation with $\sigma_1 = 3·314$ Å, or

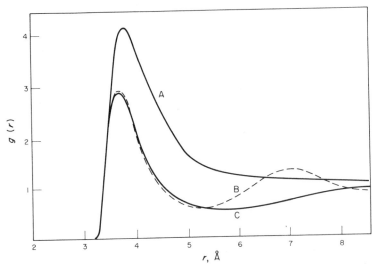

FIG. 5.3. Illustration of components of $g(r)$: curve A is the contribution of the force between atoms 1 and 2 only; curve B shows the complete $g(r)$ (parameters apply to argon at 84·5°K and 1 atm); and curve C shows the effect shown in Fig. 5.2 added to A.

just less than the hard-core radius (3·405 Å), as might be expected on physical grounds. Also shown in the Figure are the known $g(r)$ for argon and the value of $g(r)$ for $W_s(r) = 0$. It can be seen immediately that $W_s(r)$ acts to reduce $|u(r)|$ at all r over which $g(r)$ is significant, and that this effect tends to localize the second atoms (i.e. to make the first peak in $g(r)$ sharper). However when $r > 2\sigma_1$, equation (5.23) must fail badly because the third atom can fit between the first two. This occurs with a high probability, owing to the pressure of other atoms and because of the attractive part of $u(r)$. Thus an additional term must be added to $W_s(r)$ at this value of r which acts in the opposite sense to equation (5.25b) so creating the second and further peaks in $g(r)$ (Fig. 5.3). In this way it can be seen that the action of the third atoms is to isolate the second atoms and form them into a "cell" around the first atom. The extent to which this occurs depends upon the density as shown by the factor ρ in $W(r)$. At high densities, it may be argued that the cell is sufficiently well formed to justify the concept of a "cell model", in which the first atom is confined to its cell and moves more or less freely within it. However only a detailed calculation of $W(r)$ can show if this assumption is acceptable [i.e. if the cell model $W(r)$ is similar to the actual $W(r)$].

5.8. Pair Distribution Equation from Cell Theory

Although the equations of Sections 5.2 to 5.4 are fairly general, it was necessary in the end to resort to a model which was "mathematical" in the sense that it was chosen to allow the equations to be solved. To this extent the results obtained were arbitrary. However in each case the final expressions were valid in the low density limit but required testing in the high density limit. It is useful to contrast this procedure with the cell theory, since the latter may be valid at high density. In the early cell model of Lennard–Jones and Devonshire[19] an approximation was made for the total potential energy (U), by making the following assumptions—

(i) the total volume was divided into cells one for each molecule,
(ii) the cells were chosen so that their centres formed a regular lattice,
(iii) each molecule was allowed to move within its cell, independently of the others.

On the basis of this model, the potential energy is seen to be the sum of (a) the potential energy (U_0) when the molecules are at the centres of their cells and (b) the change in the potential energy when each molecule moves from the centre to a point in the cell while all the other molecules remain at their centres. Thus—

$$U\{N\} = U_0 + \sum_{i=1}^{N} [\psi(r_i) - \psi(0)] \tag{5.26}$$

where r_i denotes the position relative to the cell centre and $\psi(r)$ is the potential within a cell.

In terms of the pair potential defined at equation (2.17), the cell potential $\psi(r)$ is obtained in the following way. If r_{ij0} is the distance between cell centre j and molecule i and r_{i0j0} the distance between cell centres i and j, then the following identity holds—

$$u(r_{ij}) = u(r_{i0j0}) + [u(r_{ij0}) - u(r_{i0j0})] + [u(r_{i0j}) - u(r_{i0j0})] + \Delta_{ij} \tag{5.27}$$

where—

$$\Delta_{ij} = u(r_{ij}) - u(r_{ij0}) - u(r_{i0j}) + u(r_{i0j0})$$

Substitution of equation (5.27) into equation (2.17) gives—

$$U\{N\} = \sum_{i \le j} u(r_{i0j0}) + \sum_i \sum_j [u(r_{ij0}) - u(r_{i0j0})] + \sum_{i \le j} \Delta_{ij} \tag{5.28}$$

which is equivalent to equation (5.26) if the term in Δ_{ij} is zero. The sum over the Δ_{ij} is zero if the molecules move independently; in practice the correlations between the motions of different molecules mean that it is not zero, although it may be small compared to the other terms. It is this step that makes the problem mathematically amenable and allows $g(r)$ to be calculated.

Thus the substitution of equation (5.28)—with the last term neglected— in the definition (equation 2.2 or equation 2.6) of $n^{(2)}(r)$ leads to a relation between $g(r)$ and $u(r)$. This relation will of course, be very different from those given by the equations discussed in Sections (5.2) to (5.4). However it has the merit of containing features of the solid state rather than the gaseous state, and illustrates the significance of the initial choice of a model. In recent years this model has not been used extensively, although there have been many extensions of it.[19] In all cases, however, there is a tendency to over emphasize the solid-state aspects so indicating that Δ should not be neglected and molecules should pass through the cell walls.

5.9. Discussion of $g(r)$–$u(r)$ Equations

The preceding Sections have covered in outline four methods of relating $g(r)$ and $u(r)$. Each of the methods involves an approximation of either a physical or a mathematical character. In the case of the first three approximations, correct results are obtained in the limit of low density (the gas limit) and the high-density results are to a greater or lesser extent inaccurate. In the case of the cell theory, correct results are obtained in the high-density limit if the atoms vibrate as independent Einstein oscillators. Clearly many alterative[19] solid-state models are possible, but the ones examined so far are correct only in the limit of high density and for specified vibrational properties.

At the present stage, there is no theoretically satisfactory method of connecting $g(r)$ and $u(r)$ for a liquid. For methods in which a physical approximation is used it is possible, in principle, to test them experimentally and so gain a deeper physical insight into the liquid state. Those methods which involve a mathematical approximation can be improved by higher-order mathematical approximations.[5] Unfortunately the mathematical difficulties rapidly increase, so that it is doubtful whether substantial progress can be made by this approach. It would be convenient if all the methods could be compared on a common basis which had a clear physical significance. So far there is no simple way of doing this. Possibly the most direct method is through a comparison of the integral quantity $W(r)$—equation (5.6)— but because the full significance of such a comparison is not yet clear it is not certain that it would be useful.

The comparison of theoretical and experimental results will be postponed until Chapter 7. However, the conclusion of that discussion may be anticipated, namely that significant improvements in the experimental data and their theoretical analysis are required in order to subject the results given in this Chapter to a rigorous test.

Symbols for Chapter 5

$E(r)$	Function in BG equation, defined at equation (5.8)	t	Parameter
$F(t)$ $G(t)$	Functions defined by equation (5.19)	U_0	Equilibrium potential in cell theory
$g_1(r)$	Function defined by equation (5.14)	$\alpha, \beta, \gamma, \delta$	Coefficients in equation (5.24)
$p(\mathbf{r}_1, \mathbf{r}_2, \mathbf{r}_3)$	Probability of a third molecule being at $\mathbf{r}_3$ if two molecules are at $\mathbf{r}_1$ and $\mathbf{r}_2$	Δ_{ij}	Quantity (defined at equation 5.28) related to correlation of motion of neighbouring atoms
r_{ij0}	Distance between cell centre j and molecule i in cell theory	η	Quantity proportional to density in hard-sphere system
r_{i0j0}	Distance between cell centres	$\tau(r)$	Function defined by equation (5.15)
s	Parameter	$\psi(r)$	Potential across a cell

CHAPTER 6

Measurement of the Pair Distribution Function

6.1. Diffraction of Radiation

The scattering of radiation by condensed matter will involve the distribution of atomic positions [i.e. $g(r)$] if the wavelength of the radiation is of the order of magnitude of the interatomic spacing. If a beam of radiation falls on a target and wavelets scattered by different atoms have similar amplitudes and phases, then the scattered waves will interfere and the target is acting as a diffraction grating. In this case the distribution of scattered intensity contains information on the distribution of atoms. Almost any kind of radiation may be used for such experiments—electromagnetic (X-rays, γ-rays, light); electrons; neutrons etc. Differences arise due to the differences in the scattering properties of single atoms for each kind of radiation. These differences are summarized in the Table below—

Radiation	Scattering centre	Size of scattering centre relative to size of atom
Electromagnetic	Electron Density	1
Electrons	Charge Density	1
Neutrons	Nucleus	0

This Table shows that neutrons have simpler scattering properties than the other cases since they are essentially scattered by a point (the nucleus). In the discussion below this case will be considered first.

6.2. Neutron Scattering by a Single Atom

Consider the scattering cross section for a neutron by a free atom. In this case the wave equation is written (Schiff,[6] equation 18.8)—

$$(\nabla^2 + k^2)\psi(r) = \frac{2m}{\hbar} V(r)\psi(r) \qquad (6.1)$$

where k is the wave number of the neutron in the centre of mass system $\psi(r)$ is the wave function for the neutron and $V(r)$ is the scattering potential. The reduced mass, m, is defined by—

$$m = \frac{M}{A+1} \tag{6.2}$$

where M is the mass of the atom and A is the ratio of atomic to neutron masses. The solution to equation (6.1) which has the correct boundary conditions may be shown (Schiff,[6] equation 26.16) to be—

$$\psi(r) = e^{i\mathbf{k}_0 \cdot \mathbf{r}} - \frac{m}{2\pi\hbar^2} \int \frac{\exp ik|\mathbf{r} - \mathbf{r}'|}{|\mathbf{r} - \mathbf{r}'|} V(r')\psi(r')\, d\mathbf{r}' \tag{6.3}$$

where $\mathbf{k}_0$ is the wave vector of the incident neutron.

To proceed further, it is necessary to make an approximation for $\psi(r')$ in the second term on the right-hand side of this equation. Fermi suggested using the Born approximation in situations where the neutron wavelength and interatomic distances are much greater than the nuclear scattering length.* This approximation is to replace $\psi(r')$ by $\exp(i\mathbf{k}_0 \cdot \mathbf{r}')$ in equation (6.3), and if this is done, the $r \to \infty$ limit of the equation becomes—

$$\psi(r)_{r \to \infty} = e^{i\mathbf{k}_0 \cdot \mathbf{r}} - \frac{e^{ikr}}{r}\left[\frac{m}{2\pi\hbar^2} \int e^{i(\mathbf{k}_0 - \mathbf{k}) \cdot \mathbf{r}'} V(r')\, d\mathbf{r}' \right] \tag{6.4}$$

Thus the scattering amplitude, obtained from a comparison of equations (3.22) and (6.4), is—

$$f(\theta) = -\frac{m}{2\pi\hbar^2} \int e^{i(\mathbf{k}_0 - \mathbf{k}) \cdot \mathbf{r}'} V(r')\, d\mathbf{r}' \tag{6.5}$$

in agreement with the result obtained in Chapter 3 (equation 3.49). The scattering angle, θ is given by—

$$\cos \theta = \mathbf{k}_0 \cdot \mathbf{k}/k_0 k \tag{6.6}$$

The condition for the validity of the Born approximation is that the integral term in equation (6.3) should be small compared to $\exp i\mathbf{k}_0 \cdot \mathbf{r}$, which reduces[6,7] to the condition given in Chapter 3. It is easily shown that for neutron energies less than 1 keV this condition is far from satisfied, so that the magnitude of $f(\theta)$ cannot be calculated from (6.5). However since the scattering length is much less than the neutron wavelength, the volume over which the wave function differs

---

* The "scattering length" will be defined later; it is of the order of magnitude of the nuclear radius.

significantly from the plane wave value is negligible compared to the total volume, $f(\theta)$ will be given correctly by the Born approximation. Because the nucleus is almost a point scatterer, Fermi approximated $V(r)$ by—

$$V(r) \simeq \frac{2\pi\hbar^2}{m} a\delta(r) \qquad (6.7)$$

where the constant a is called the "scattering length". The error in the approximation (6.7) is roughly the ratio of the scattering length to the neutron wavelength or $\sim 10^{-4}$ for a 1 Å neutron. It is easily checked that the substitution of equation (6.7) in equation (6.5) gives $f(\theta) = -a$, or a free atom scattering cross-section of $4\pi a^2$. The problem is solved apart from a constant, a, which can be determined experimentally for each nucleus. For convenience, a scattering length (b) for a "bound atom" is normally used in equation (6.7), in order to introduce the actual neutron mass m_n in place of the reduced mass m; thus—

$$\frac{b}{m_n} = \frac{a}{m} \quad \text{or} \quad b = a\left(\frac{A+1}{A}\right) \qquad (6.8)$$

The above discussion implied that there was only one nuclear scattering length, a. But in general the nucleus will have a non-zero spin, I, and the partial wave analysis[6,7] of the scattering shows that because of the small angular momentum of the neutron for wavelengths ~ 1 Å only s waves contribute. Thus there are two possible spin states of the compound nucleus ($I+\frac{1}{2}$ and $I-\frac{1}{2}$). By definition the s wave scattering is isotropic and energy independent, and these properties are generally valid for neutrons having energies less than 1 keV. Now in cases where a virtual nuclear level is close, the scattering lengths for the two spin states will be different,* and are denoted by a_+ and a_-. The $+$ state carries a statistical weighting factor of $(I+1)/(2I+1)$, and the $-$ state carries a weighting factor of $I/(2I+1)$. Thus the free atom cross-section for unpolarized neutrons will be—

$$\sigma_{fa} = 4\pi\left\{\frac{I+1}{2I+1}a_+^2 + \frac{I}{2I+1}a_-^2\right\} \qquad (6.9)$$

In what follows the interference of waves scattered by different nuclei will be considered. The scattering length (a_{coh}) for this effect

---

* Note that for neutron wavelengths ~ 1 Å, the neutron energy is far too small to excite the internal nuclear levels. Thus the only possible processes are nuclear absorption and elastic scattering.

is given by the sum of the scattering lengths for the two spin states, i.e.—

$$a_{coh} = \frac{I+1}{2I+1}a_+ + \frac{I}{2I+1}a_-$$ (6.10)

Clearly the cross-section $4\pi a_{coh}^2$ is less than σ_{fa} given by equation (6.9), and the difference is called the incoherent cross-section, or—

$$\sigma_{incoh} = \sigma_{fa} - 4\pi a_{coh}^2 = 4\pi(a_+ - a_-)^2 \frac{I(I+1)}{(2I+1)^2}$$

The incoherent scattering length obtained from this equation is—

$$a_{incoh} = \frac{\sqrt{I(I+1)}}{2I+1}(a_+ - a_-)$$ (6.11)

and the definitions of b_{coh} and b_{incoh} follow from equation (6.8). Equation (6.11) defines an incoherent scattering length which determines the magnitude of the scattering for which interference does not occur—i.e. the scattered wave from each atom escapes from the target without modification.

If the target is an element that contains several isotopes (each having a scattering length different from the others) which are randomly distributed, the scattering will be divided into coherent and incoherent components in a similar way to that described above. Since the incoherent scattering corresponds to scattering by the atoms taken singly, it is isotropic. For this reason, it constitutes a background that has to be measured and subtracted from the total scattering. For the remainder of this Chapter, only the coherent scattering will be considered. Discussion of incoherent scattering will be resumed in Chapter 8.

6.3. The measurement of $g(r)$ by Neutron Scattering

The coherent scattering cross-section of the assembly of nuclei in a liquid is now evaluated readily from equation (6.5). An interaction potential $V_L(r')$ for the whole system is required, and this is given by a sum of terms of the Fermi type (equation 6.7), i.e.—

$$V_L(r') = \frac{2\pi\hbar^2}{m_n} \frac{b_{coh}}{N} \sum_i \delta(r' - r_i)$$ (6.12)

where r_i is the position of the ith nucleus. Thus combining equations

(3.23), (6.5) and (6.12), the cross-section is—

$$\left(\frac{d\sigma}{d\Omega}\right)_n \equiv \sigma(\theta) = \langle f^*(\theta)f(\theta)\rangle = b_{coh}^2 \frac{1}{N}\left\langle \sum_{i \leq j} e^{i\mathbf{Q}.\mathbf{r}_{ij}}\right\rangle \qquad (6.13)$$

where—

$$\mathbf{Q} = \mathbf{k}_0 - \mathbf{k}$$

Since a cross-section is measured by counting the number of neutrons scattered during a period of time long compared to the lifetime of a thermal fluctuation, the quantity measured is a thermal average denoted by $\langle---\rangle$.

In writing equation (6.13) it has been assumed that $\mathbf{r}_i$ and $\mathbf{r}_j$ are the positions of atoms i and j at the same moment of time. For this to be possible, the recoil in the neutron scattering process must be negligible, and this implies for example, that $m_n = m$. An alternative condition will be derived in Chapter 8. Equation (6.13) may be rewritten—

$$\left(\frac{d\sigma}{d\Omega}\right)_n = b_{coh}^2 \int e^{i\mathbf{Q}.\mathbf{r}}\left\langle \frac{1}{N}\sum_{ij}\delta(\mathbf{r}+\mathbf{r}_i-\mathbf{r}_j)\right\rangle d\mathbf{r}$$

$$= b_{coh}^2\left\{\int e^{i\mathbf{Q}.\mathbf{r}}\,d\mathbf{r}\rho[g(r)-1]+1+\rho\delta(\mathbf{Q})\right\} \qquad (6.14)$$

where—

$$\rho g(r) = \frac{1}{N}\left\langle \sum_{i \neq j}\delta(\mathbf{r}+\mathbf{r}_i-\mathbf{r}_j)\right\rangle \qquad (6.15)$$

The definition (6.15) excludes the case $i = j$ in equation (6.13); this case leads to the term $+1$ in equation (6.14). It is easy to see that the physical meaning of equation (6.15) is the same as the physical definition of $n^{(2)}$ in Chapter 2, namely equation (6.15) is the probability of finding an atom at $\mathbf{r}_j$ if there is an atom at $\mathbf{r}_i$. Thus $g(r)$ at equation (6.14) is the same quantity as at equations (2.9a and b) and consequently equation (6.14) may be used to derive $g(r)$ from measurements of neutron scattering cross-sections. A schematic layout of such an experiment is given in Fig. 6.1.

6.4. The Measurement of $g(r)$ by X-ray or Electron Scattering

The scattering of X-rays by a free atom involves the excitation of atomic energy levels rather than the nuclear energy levels discussed in the case of neutron scattering. However, unlike the neutron case,

X-rays of 1 Å wavelength have sufficient energy to excite many atomic levels, and this leads to inelastic processes known as "fluorescent" and "Compton" scattering. Since $g(r)$ is (simply) related to the elastic cross-section only, it is necessary for the present purpose to calculate or measure these inelastic cross-sections and subtract them from the total cross-section in order to find the elastic cross-section. The details of this step will not be discussed here. Similar remarks apply to electron scattering. Also because the waves scattered by each atom have the same amplitude and phase the scattering is coherent.

The scattering potential is small compared to the energy of the radiation, and thus the Born approximation may be used for both X-ray and electron scattering (although since the long-range Coulomb interaction is important in the electron case, the Born approximation is satisfactory only for high-energy electrons). However unlike the neutron case, the scattering lengths may be calculated from first principles—

(i) Scattering by an isolated electron gives

$$b_x = \frac{e^2}{m_e c^2} \quad \text{(Schiff,[6] Section 23)} \tag{6.16a}$$

(ii) Scattering by a charge, e, gives

$$b_e = \frac{2 m_e e^2}{\hbar^2 Q^2} \quad \text{(Schiff,[6] Section 20)} \tag{6.16b}$$

where m_e denotes the electron mass and c the velocity of light. Since X-ray scattering involves the virtual excitation of many levels, the electrons are treated as free particles. The factor $1/Q^2$ arises in the case of electron scattering from the range of the Coulomb potential (i.e. b_e is proportional to the Fourier transform of the potential).

Following the method used in Section 6.3, the cross-sections are respectively—

$$\left(\frac{d\sigma}{d\Omega}\right)_x = \left(\frac{b_x}{e}\right)^2 \int e^{i\mathbf{Q}\cdot\mathbf{r}} G_x(r)\, d\mathbf{r} \tag{6.17a}$$

and—

$$\left(\frac{d\sigma}{d\Omega}\right)_e = \left(\frac{b_e}{e}\right)^2 \int e^{i\mathbf{Q}\cdot\mathbf{r}} G_e(r)\, d\mathbf{r} \tag{6.17b}$$

In the case of X-rays, there is no change of polarization for the component of the scattering which is related to $G_x(r)$. The correlation functions G_x and G_e are written in this case in quantum mechanical form[20]—

$$G_x(r) = \frac{1}{N} \int d\mathbf{r}' \langle \rho_x(\mathbf{r}') \rho_x(\mathbf{r}' + \mathbf{r}) \rangle \tag{6.18a}$$

where $\rho_x(\mathbf{r})$ is the electron-density $\rho_x(\mathbf{r}) = -e\sum_l \delta(\mathbf{r} - \mathbf{r}_l)$, and—

$$G_e(r) = \frac{1}{N} \int d\mathbf{r}' \langle \rho_e(\mathbf{r}')\rho_e(\mathbf{r} + \mathbf{r}')\rangle \qquad (6.18b)$$

where $\rho_e(\mathbf{r}')$ is the charge-density operator $\rho_e(\mathbf{r}) = \Sigma_l e_l \delta(\mathbf{r} - \mathbf{r}_l)$. The summation extends over all the electrons with charge $e_l = -e$ and nuclei with charge $e_l = Ze$ (where Z is the atomic number). In the above equations $\mathbf{r}_l$ denotes the position operator for electrons or charges.

To proceed further, equations 6.18(a) and (b) must be expressed in terms of the distribution of nuclei. This may be done through the electron distribution (ρ_x^0) around each nucleus, which is defined by—

$$\rho_x(\mathbf{r}) = \sum_i \rho_{xi}^0(\mathbf{r} - \mathbf{r}_i) \qquad (6.19)$$

where $\mathbf{r}_i$ is the nuclear position, and similarly for the charge distribution. Now define a form factor as—

$$F_x(\mathbf{Q}) = \frac{1}{e} \int e^{i\mathbf{Q}\cdot\mathbf{r}} \langle \rho_x^0(\mathbf{r})\rangle \, d\mathbf{r} \qquad (6.20)$$

and the cross-section may be expressed in terms of the nuclear

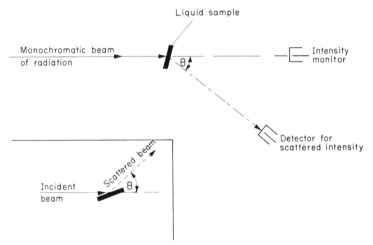

FIG. 6.1. Schematic layout for the measurement of $d\sigma/d\Omega$ (equations 6.14 and 6.21). The ratio of the intensity recorded by the detector of scattered radiation to that recorded by the intensity monitor, is converted to $d\sigma/d\Omega$ after corrections for unwanted effects. The sample is contained in a vessel and mounted in a furnace or cryostat to control the temperature at the required value. Sometimes the sample is pressurized. An alternative scattering geometry is shown in the inset.

positions and $F(Q)$. On combining equations (6.13), (6.17), (6.18), (6.19) and (6.20), it is found for scattering without a change of polarization, that—

$$\frac{1}{b_x^2}\left(\frac{d\sigma}{d\Omega}\right)_x = |F_x(Q)|^2 \left(\frac{d\sigma}{d\Omega}\right)_n \frac{1}{b_{coh}^2} \qquad (6.21)$$

and similarly for the electron case.

Thus if the function $F(Q)$ is known then $g(r)$ may be calculated from the X-ray or electron elastic cross-sections. It is usual to assume that $F(Q)$ is approximately equal to the value calculated for a free atom, since the major contribution to $F(Q)$ comes from the core electrons whose orbits are unchanged when free atoms are condensed into a liquid. For accurate work corrections to the free-atom model are required, or, alternatively, the corrections may be determined experimentally via equation (6.21) and the neutron cross-sections. A schematic layout of a scattering cross-section experiment is shown in Fig. 6.1.

6.5. The Structure Factor $S(Q)$

The structure factor $S(Q)$ is defined by—

$$S(Q) = 1 + \rho \int_V e^{iQ \cdot r}[g(r) - 1] \, dr \qquad (6.22)$$

and for $Q \neq 0$ it is proportional to the neutron cross-section (equation 6.14). Its general shape is shown in Fig. 6.2a; in a liquid it is a function only of the magnitude of Q, since rotation of the target does not alter the intensity of scattered radiation. For liquids near the triple point, the value of $S(Q)$ is much less than unity (the actual value depends on χ_T via 6.23), and the curve is dominated by a large peak at a value of $Q \sim 2\pi/r_0$ where r_0 (Fig. 2.1) is approximately the atomic spacing. At higher values of Q, $S(Q)$ oscillates and for $Q \to \infty$ the oscillations damp completely. In practice, it is found that detectable oscillations extend to $Q \sim 20 \, \text{Å}^{-1}$, and the major peak is found at $Q \sim 2 \, \text{Å}^{-1}$.

The value of $S(0)$ is readily obtained from equation (2.29) as—

$$S(0) = \rho k T \chi_T \qquad (6.23)$$

and consequently measurement of $S(0)$ is equivalent to determining the equation of state. At high temperatures, approaching the critical point, χ_T is large and thus $S(0) \gg 1$. In this region, $S(0)$ is the maximum value of $S(Q)$ and the low Q section dominates the behaviour of the whole curve.

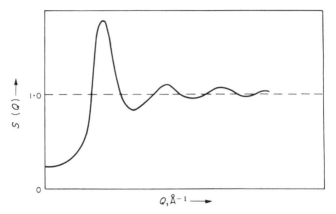

FIG. 6.2a. Typical shape of the function $S(Q)$ for the liquid region shown in Fig. 1.1. The limiting value for $Q \to 0$ is given by equation (6.23).

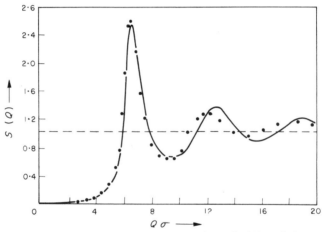

FIG. 6.2b. Comparison of hard sphere calculation of $S(Q)$ and the experimental values for Rb at 40°C. The full line is the calculated curve and the full circles are the experimental points; the value of σ was 4.30 Å and the actual liquid density was used in the calculation. (From Ashcroft and Leckner.[21])

The Ornstein–Zernike equation (5.9) may be Fourier-transformed to give—

$$h(Q) = c(Q) + \rho c(Q) h(Q) \qquad (6.24a)$$

or—

$$S(Q) - 1 = \rho h(Q) = \frac{\rho c(Q)}{1 - \rho c(Q)} \qquad (6.24b)$$

Thus a calculation of $c(Q)$ is readily converted into $S(Q)$. For example the PY hard-sphere expression equation (5.26) is easily transformed to give—

$$\rho f_{ph}(Q) = -\frac{24\eta}{(1-\eta)^4} \int_0^1 ds \, s^2 \frac{\sin sQ\sigma}{sQ\sigma}$$

$$\times \left[(1+2\eta)^2 - 6\eta \left(1 + \frac{\eta}{2}\right)^2 s + \frac{\eta}{2}(1+2\eta)^2 s^3 \right] \qquad (6.25)$$

Ashcroft and Leckner[21] have compared this expression to the measured $S(Q)$ for several liquid metals. If the ion–ion interaction is the dominant feature in $u(r)$ then a hard-sphere approximation may be reasonable. Figure 6.2(b) shows an example of this comparison for Rubidium at 40°C. The principal features of the curve are reproduced by equation (6.25) for reasonable values of the density and hard-sphere radius. Work of this kind indicates that the main features of $g(r)$ and $S(Q)$ are fixed by geometrical packing considerations (which are controlled by the core of $u(r)$—see Fig. 2.2 and Bernal[8]), and that the tail of $u(r)$ at large r controls the detailed behaviour of $S(Q)$.

6.6 $S(Q)$ for Molecular Liquids

The discussion given above has referred to a monatomic liquid. In the case of a molecular liquid the quantity required is $g(r_c)$ where r_c denotes the position of a molecular centre rather than a particular atom. In terms of equation (6.13), the structure factor deduced from the cross-section for a molecular liquid may be written—

$$S_m(Q) = \frac{1}{N(\Sigma b_n)^2} \left\langle \left| \sum_i \sum_n b_n \, e^{iQ \cdot r_{ni}} \right|^2 \right\rangle \qquad (6.26)$$

where b_n is the scattering length of the nth atom in the molecule and r_{ni} is the position of the nth atom in the ith molecule. A "centre" of the ith molecule may be defined by—

$$(r_c)_i = \frac{\Sigma b_n r_{ni}}{\Sigma b_n} \qquad (6.27)$$

(Note, for a random molecular orientation the distribution of r_c is independent of the values of b_n).

Then equation (6.26) may be rewritten in terms of r_c as—

$$S_m(Q) = \frac{1}{N(\Sigma b_n)^2} \left\langle \sum e^{iQ \cdot r_{cij}} \left[\sum_n b_n \, e^{iQ(r_{cni} - r_{cnj})} \right] \right\rangle \qquad (6.28)$$

where r_{cij} is the vector between the centres of the ith and jth molecules, and r_{cni} is the vector between the centre of the ith molecule and the atom n (i.e. $r_{ni} = r_c + r_{cni}$).

At this stage, it is necessary to specify the correlations between the atoms in a molecule and also the correlations between the orientations of different molecules. Equation (6.28) may be rewritten as—

$$S_m(Q) = f_1(Q) + \frac{1}{N} \left\langle \sum_{i \neq j} e^{iQ.r_{cij}} \right\rangle f_2(Q) \qquad (6.29)$$

where—

$$f_1(Q) = \frac{1}{(\Sigma b_n)^2} \left\langle \left| \sum_n b_n e^{iQ.r_{cn}} \right|^2 \right\rangle$$

and $f_2(Q)$ is a function of the correlation factors mentioned above. In the particular case (to which the theory of Chapter 2 and 3 applies) where the molecules are randomly oriented the function $f_2(Q)$ may be written—

$$f_2(Q) = \frac{1}{(\Sigma b_n)^2} \left[\sum b_n \frac{\sin Qr_{cn}}{Qr_{cn}} \right]^2 \qquad (6.30)$$

For Q small, both $f_1(Q)$ and $f_2(Q)$ approach unity, in which case equation (6.29) reduces to—

$$S(Q) = 1 + \frac{1}{N} \left\langle \sum_{i \neq j} e^{iQ.r_{cij}} \right\rangle \qquad (6.31a)$$

Use of $S(Q)$ in (6.22) then gives $g(r_c)$ as required. For larger values of Q the details of the molecular structure and orientation must be known and employed to determine $f_1(Q)$ and $f_2(Q)$. Once $f_1(Q)$ and $f_2(Q)$ are known, $S(Q)$ can be constructed from $S_m(Q)$ through the relation [deduced from equations 6.29 and 6.31a]—

$$S(Q) = 1 + \frac{S_m(Q) - f_1(Q)}{f_2(Q)} \qquad (6.31b)$$

Thus the neutron scattering cross-sections of a molecular liquid do not lead directly to the required function [$S(Q)$], and it is necessary to interpret the data via equation (6.31b). For X-ray scattering, this problem does not have a rigorous solution, and various approximate solutions have been suggested. The most widely used is to write equation (6.20) as—

$$F_{xn}(Q) \sim Z_n F_{x0}(Q) \qquad (6.32)$$

where $F_{x0}(Q)$ is an average form factor (per electron) for the atoms in the liquid and Z_n is the number of electrons in the nth atom. In this case the "known" value of $|F_{x0}(Q)|^2$ may be divided out and the analysis may proceed along lines similar to those above.

Symbols for Chapter 6

$a_\pm$	Scattering lengths for $I \pm \frac{1}{2}$ spin states	I	Spin of target nucleus
a_{coh}	Coherent scattering length	Z	Atomic number
a_{incoh}	Incoherent scattering length	ρ_e	Charge density
$f(\theta)$	Scattering amplitude	ρ_x	Electron density
$f_1(Q)$ $f_2(Q)$	Form factors for molecular scattering		

CHAPTER 7

Discussion of Equilibrium Properties

So far it has been shown how the thermodynamic properties $(P, V, T, C_V$ etc.) of a liquid may be expressed in terms of two functions —the pair potential $u(r)$, assumed independent of P, V, T etc. and the pair distribution function $g(r)$, which is a function of density and temperature. The methods by which these two functions may be calculated from first principles or measured in scattering experiments have been discussed. None of these methods is free from theoretical or experimental error, so that there is some uncertainty in the derived values of the thermodynamic quantities.

In this Chapter the relations between $g(r)$ and $u(r)$ will be used to compare the data on the two functions. This gives some indication of the uncertainties in the present theory, as well as some idea of where it can be successful. Then a discussion of the equations of state for dilute and dense gases is given to show that the methods of calculation discussed earlier are fairly successful. However in the liquid range (shown by shading in Fig. 1.1) serious problems arise and the same methods are much less satisfactory. These problems are discussed in Section 7.6 and the current position is summarized at the end of the Chapter.

7.1. Evaluation of $g(r)$ from $u(r)$

If it were possible to calculate $g(r)$ from $u(r)$ with high accuracy, the thermodynamic properties could be readily calculated from first principles. Thus the first task is to examine this possibility. As shown by Fig. 6.2, a tolerably good fit to the function $S(Q)$ [and hence to $g(r)$] can be obtained even when a crude form for $u(r)$ is used. This case might imply that the thermodynamic properties of a hard-sphere fluid should approximate those of a metal, but, as shown later, this is not so. Thus in this example reliable data would not be obtained from such a simple calculation.

Another example is liquid argon for which Khan[22] has calculated $g(r)$ using the Lennard–Jones potential (Chapter 3) and the PY and

HNC equations (Chapter 5). A comparison of calculation with experiment is given in Fig. 7.1(a). Two features are apparent from this result, namely, the two theories are not in good agreement and both theories disagree with the experimental data. The PY theory fits the experimental data somewhat better than the HNC theory. Figure 7.1(b) shows a comparison of the cell model for (L–J) argon with experiment; although some experimental features are reproduced the detailed

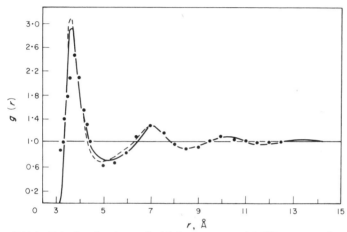

FIG. 7.1(a). Calculated values of $g(r)$ for argon at 84·4°K compared to neutron-diffraction data. The full line shows the HNC calculation, and the dotted line shows the P–Y calculation, both using the L–J potential (from Kahn[22]). The full circles indicate the experimental points (from G. D. Henshaw, *Phys. Rev.*, 1957, **105**, 976).

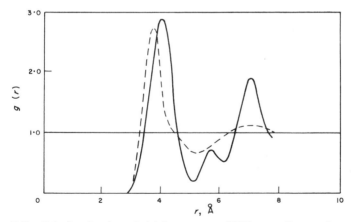

FIG. 7.1b. Calculated value of $g(r)$ for argon at 90°K according to the cell theory (full line) compared to the X-ray diffraction data (dotted line) (from Barker,[19] Fig. 10.1).

agreement is worse than in Fig. 7.1(a). The extent of the agreement between theory and experiment shown in Figs. 7.1(a) and (b) (as well as in Fig. 6.2) is indicative of the success of these methods. They give the correct form and magnitude of $g(r)$, but fail to give fine accuracy. In particular the failure to give a close fit at low r should be noted.

One method of checking the results obtained from the $[g(r)-u(r)]$ equations of Chapter 5 is provided by the density and temperature derivatives of $g(r)$. The temperature derivatives are complicated, the lowest derivative of $n^{(2)}$ involves $n^{(4)}$, which has not been employed in the theory up to this point. However the lowest density derivative of $n^{(2)}$ involves only $n^{(3)}$; it may be obtained from the general definition of Chapter 2 by straightforward differentiation, as shown by Schofield.[23] The result is—

$$\left(\frac{\partial \rho^2 g(r)}{\partial \rho}\right)_T = \frac{2\rho g(r) + \rho^2 \int [g^{(3)}(\mathbf{r}, \mathbf{s}) - g(r)]\, ds}{1 + \rho \int [g(r) - 1]\, dr} \tag{7.1}$$

where $\mathbf{r}$ and $\mathbf{s}$ are as defined in Fig. 5.1. If the same approximation for $n^{(3)}$ is used in both equations (5.6) and (7.1), then equation (7.1) may be used to test the reliability of the calculated $g(r)$ at each value of r. It should be noted that if both equations (5.6) and (7.1) give consistant data the pressure and compressibility equations of state will be consistant also.[23] A detailed test of this kind is not available, so far, for the cases quoted above (the form of $\partial g/\partial \rho$ is shown in Fig. 7.4h).

7.2. Evaluation of $u(r)$ from $g(r)$

An alternative use of the equations discussed in Chapter 5 is to employ them to calculate $u(r)$ from a measurement of $g(r)$. This is particularly useful in the case of metals, since both the theoretical derivation and the experimental tests of $u(r)$ are less satisfactory than in the case of the non-conducting liquids. For the PY and HNC equations, $u(r)$ may be obtained fairly directly from the measurements of $S(Q)$. First $h(Q)$ and $c(Q)$ are derived via equation (6.24b) and then they are Fourier-transformed to give $h(r)$ and $c(r)$. Finally equation (5.13a) is used to give—

$$\left. \begin{array}{l} u(r)_{\text{HNC}} = kT[h(r) - \ln\{1 + h(r)\} - c(r)] \\[2mm] u(r)_{\text{PY}} = kT \ln\left[1 - \dfrac{c(r)}{1 + h(r)}\right] \end{array} \right\} \tag{7.2}$$

The YBG equation involves a more complex analysis to find $u(r)$.

Johnson *et al.*[24] have derived $u(r)$ from $g(r)$ by using the YBG and PY equations. The cases of sodium and argon are shown in Fig. 7.2. It can be seen that the two theories do not agree satisfactorily, but that the long-range oscillations expected for a metal are found in both the YBG and PY results. However, the numerical difficulties in work of this kind may have caused the oscillations of Fig. 7.2(a) to be exaggerated.[13] In the case of argon (Fig. 7.2b) the general shape of the potential is in satisfactory agreement with the predictions of Chapter 3,

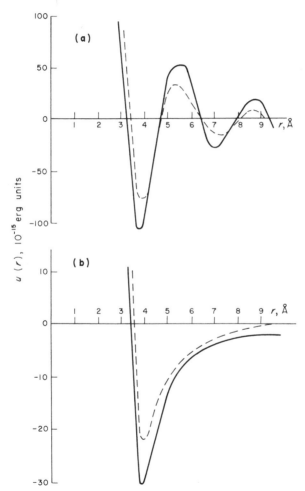

FIG. 7.2. Potential functions $u(r)$ derived from measurements of $g(r)$. The full lines indicate results obtained from the YBG equation, and the dotted lines indicate results from the PY equation (from Johnson *et al.*[24]). (a), sodium at 203°C; (b), argon at 84°K.

but neither the long-range nor short-range parts of $u(r)$ are obtained accurately from the present data. Broadly speaking the potential well is derived by this method, rather than the features examined in Chapter 3. In this sense, the above method is complementary to those of Chapter 3.

7.3. The Equation of State for a Dilute Gas

The compressibility equation of state may be written in terms of the direct correlation function as follows. Equation (2.29) is—

$$\left(\frac{\partial \rho}{\partial P}\right)_T kT = \rho \int_T h(r) \, d\mathbf{r} + 1 \tag{7.3a}$$

or—

$$\frac{1}{kT}\left(\frac{\partial P}{\partial \rho}\right)_T = \frac{1}{1 + \rho \int_V h(r) \, d\mathbf{r}} = 1 - \rho \int_V c(r) \, d\mathbf{r} \tag{7.3b}$$

where equation (6.24a) has been employed to make the last step. This equation should be compared to the pressure equation of state (equation 2.24)—

$$\frac{P}{kT} = \rho - \frac{\rho^2}{6kT} \int_V r \frac{du(r)}{dr} g(r) \, d\mathbf{r} \tag{7.4}$$

or—

$$\frac{1}{kT}\left(\frac{\partial P}{\partial \rho}\right)_T = 1 - \frac{\rho}{3kT} \int_V r \frac{du(r)}{dr}\left[g(r) + \frac{\rho}{2}\left(\frac{\partial g(r)}{\partial \rho}\right)_T\right] d\mathbf{r} \tag{7.5}$$

As an example, consider the $g(r)$ for a dilute gas (equation 2.33), which is independent of the density. Thus the term $\partial g/\partial \rho \to 0$ in equation (7.5) and the pressure equation of state can be rewritten in terms of $c(r)$, i.e.—

$$\frac{P}{\rho kT} = 1 - \frac{\rho}{2} \int_V c(r) \, d\mathbf{r} \tag{7.6a}$$

or—

$$\frac{P}{kT} = \frac{\rho}{2} + \frac{1}{2kT\chi_T} \tag{7.6b}$$

If $c(r)$ is approximated by its density independent first term—equation (5.11)—it is seen that equation (7.6a) reduces to the first two terms of the virial expansion given in equation (2.35). In this way (or by taking equation (7.5) to the next higher term in the density expansion) it can be seen that equation (7.6b) includes the second virial term. Hence it is a suitable equation of state for a dilute gas; that is a gas for which the density expansion can be truncated at the ρ^2 term.

In some cases the equations of state calculated from equation (7.4) and the $g(r)$–$u(r)$ relations of Chapter 5 will include equation (7.6b) as the leading term. This is true for example[25] of the HNC equation of state, and consequently at the triple point where both χ_T and P are small there will be very poor agreement between theory and experiment. The basic reason for this is that the product $(du/dr)g(r)$—equation (7.4)—depends sensitively upon the details of $g(r)$. This is illustrated by the case of argon where the two terms on the right-hand side of equation (7.4) are each 500 times larger than the left hand side at the triple point, falling to two or three times larger at the critical point.

7.4. Equation of State for PY Hard-sphere Fluid

It has been pointed out that the PY equation may be solved for a hard-sphere potential and that in some cases (Fig. 6.2) this gives a fair prediction for $g(r)$. The solution (equation 5.26) is readily employed in the pressure and compressibility equations (equations 7.3 and 7.4) to give—

$$\frac{P}{\rho kT} = \frac{1+2\eta+3\eta^2}{(1-\eta)^2} \quad \text{(pressure equation)} \tag{7.7}$$

$$= \frac{1+\eta+\eta^2}{(1-\eta)^3} \quad \text{(compressibility equation)} \tag{7.8}$$

It can be seen immediately that the pressure predicted by these equations is greater than that of the perfect gas (for a dense system in which $\eta \sim 0.5$, it is an order of magnitude larger), but in a liquid the pressure is known to be very much less than that in a gas at the same density. Consequently although the $S(Q)$ obtained by this method is fairly reasonable, the value of the pressure is quite unreasonable. An attractive term in $u(r)$ is needed to remedy this defect (Fig. 7.4).

One estimate of the range of η over which equations (7.7) and (7.8) may be used can be found from the difference between them, i.e. $3\eta^3/(1-\eta)^3$. Thus the two equations are identical as far as the η^2 term; the difference between them is negligible compared to the main term if—

$$6\eta^3 \ll 1+\eta+\eta^2 \tag{7.9}$$

For $\eta < 0.25$, this inequality is satisfied to better than a factor 10; in contrast the case quoted in Fig. 6.2 requires $\eta \sim 0.45$ which is well outside the range of validity of this method. It might be guessed that $g(r)$ would be poor at low r, so giving a poor $S(Q)$ at high Q. Examination of Fig. 6.2 shows this to be the case, although by adjusting the value of σ the agreement at high Q could be improved at the expense of the low Q region.

7.5. Comparison of Equations of State for a Dense Gas

One way in which the approximations of Sections 5.2 to 5.4 may be tested is to calculate $g(r)$ from a given $u(r)$ (by numerical methods) and then evaluate the $P-V$ relationships by the use of both equations (7.3) and (7.4). A comparison of the two results with a more accurate calculation is a useful guide to the accuracy of the data. An example of this method is shown in Fig. 7.3(a) and (b). The results were calculated for the Lennard–Jones potential using the constants for argon (Table 3.1), and the continuous line shows the result of a molecular dynamics calculation for a temperature of $2.74\varepsilon/k$. In this method, the equations of motion for all the particles in a box, filled to a chosen density, are solved continuously and the positions and motions at each instant of time are recorded. From this information the "exact" value of the pressure may be calculated. Also shown in the figure are the results obtained from equations (7.3) and (7.4) using the three approximate methods of calculating $g(r)$ discussed in Chapter 5. It can be seen that the YBG method gives poor agreement, while the HNC method is better and the PY method is in relatively good agreement with the molecular dynamics result.

A less satisfactory test is to perform the same calculation for the hard-sphere potential (equation 3.12), and in this case also an "exact" result is known from the molecular dynamics method of numerical calculation. Thus the approximate results may be compared to the known result (but note that only the value of $g(r)$ at the hard-sphere radius is needed to find the pressure). An example of this kind is shown in Fig. 7.3(c): the whole equation of state can be shown on a single

FIG. 7.3. A comparison of equations of state for dense gases. (a), Lennard–Jones potential used with the pressure equation (7.4), $T = 2.74\ \varepsilon/k$; (b), Lennard–Jones potential used with the compressibility equation (7.3), $T = 2.7\varepsilon/k$; (c), Hard sphere potential results, the letters in brackets indicate that the pressure (P) or compressibility methods (C) were used. In all three figures the continuous line represents the molecular dynamics data while the results obtained by the approximate methods are marked with the usual abbreviations. Data in (a) and (b) were taken from Levesque[63] and for (c) from Hutchinson, Thesis, Newcastle University (1963).

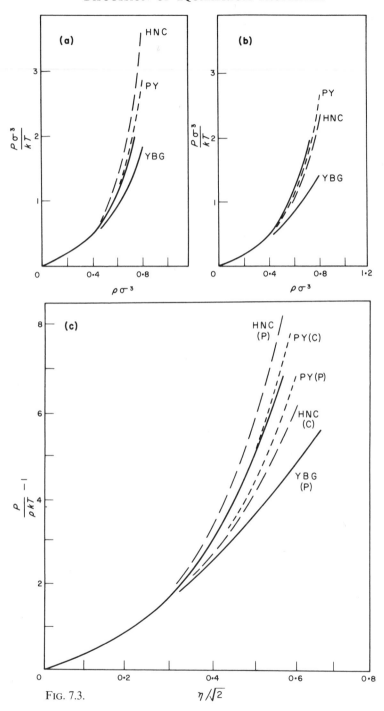

FIG. 7.3.

diagram in this case (see equations 7.7 and 8). It can be seen that the YBG equation gives the poorest result, while the HNC equation is next and the PY equation is more accurate than either of the others. This is in agreement with the conclusion from Fig. 7.3(a) and (b).

The success of the PY equation, as illustrated in Fig. 7.3, has led to its widespread use in numerical calculations of $g(r)$, although it should be borne in mind that tests such as those described above are not necessarily relevant to all problems.

7.6. Problems in the Calculation of Pressure, Energy and Specific Heat for a Liquid

The equations for pressure and energy (Chapter 2) involve the product of $u(r)$ or $\dot{u}(r)$ and $g(r)$. This product is very sensitive to the value of $g(r)$ at low values of r, since $u(r)$ has a high value there. In the limit $r \to 0$, $g(r)$ vanishes sufficiently strongly that the product vanishes. Thus the peak value is obtained for r just larger than r_m, where r_m is the largest value of r for which $g(r) \sim 0$. This behaviour is illustrated in Fig. 7.4(a–h). It is for this reason that the theoretical values of $g(r)$—Fig. 7.1—should be compared at low r. The range of these products is roughly the range of $u(r)$, and thus the long range behaviour of $g(r)$ does not affect them. Figures 7.4(d), (g) and (e) show the products appearing in the energy, specific heat and pressure integrals, respectively.

The specific heat depends upon the sum of two quantities of roughly equal size (equation 2.19)—Fig. 7.4(f) and (g). However, the variation of $g(r)$ with temperature is required, and this has not been evaluated properly because it depends upon distribution functions up to $n^{(4)}$. Approximate values of $(\partial g(r)/\partial T)_V$ may be obtained from an examination of the temperature dependence of the YBG, HNC and PY equations. From the experimental point of view, equation (2.19) may be used to calculate the specific heat from measured values of $(\partial g(r)/\partial T)_V$, and as shown in Fig. 7.4(f) this is sensitive to the minimum of $u(r)$. Detailed experimental work from which $(\partial g(r)/\partial T)_V$ could be evaluated is not available.

The pressure equation (7.4) depends upon the difference between two large numbers, and it has been commented that this formula gives inaccurate results when the values of $g(r)$ are computed by the method discussed in Section 7.1. For argon near the triple point, the difference is 0.2% of either number, so that the present knowledge of $g(r)$ (approximately 10% accurate), is inadequate. However this observation enables the value of the integral in equation (7.4) to be

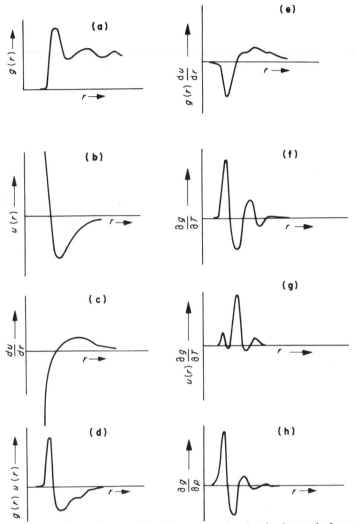

FIG. 7.4. Illustration of several functions appearing in the integrals for pressure, energy and specific heat. These curves are of qualitative significance only.

quoted as—

$$\frac{\rho}{6kT} \int_V r \frac{du(r)}{dr} g(r) \, d\mathbf{r} \simeq 1 \text{ (at the triple point)} \qquad (7.10)$$

The theoretical significance of this interesting result is not yet clear, but it is of value in estimating other properties near the triple point (equations 13.53, for example).

Since equation (7.4) is difficult to evaluate numerically, it would be convenient if equation (7.3) offered an easier route to the calculation of the pressure. Unfortunately similar difficulties occur because $(\partial\rho/\partial P)_T$ is very small near the triple point, and it is found that—

$$\rho \int_V h(r)\, d\mathbf{r} \simeq -1 \begin{Bmatrix} +5\% \text{ for argon} \\ +1\% \text{ for metals} \end{Bmatrix} \qquad (7.11a)$$

The integration over $h(r) = g(r) - 1$, involves a large cancellation of terms, since it oscillates between positive and negative values. For this reason the theoretical results for $g(r)$ lead to poor values of the compressibility (and hence for the pressure). Equations (7.10) and (7.11a) may be used in equations (7.3) and (7.5) to show that (near the triple point)—

$$\chi_T^{-1} \simeq -\rho kT - \frac{\rho^3}{6} \int_V r \frac{du(r)}{dr} \left(\frac{\partial g(r)}{\partial \rho}\right)_T d\mathbf{r} \qquad (7.11b)$$

where $\chi_T^{-1} \sim 30\rho kT$. It is possible that this equation is a better one to use for the numerical evaluation of compressibilities in the liquid state, but this point has not been tested. In any case it underlines the importance of measuring $(\partial g(r)/\partial\rho)_T$ in order to test the theory; since in some cases $du(r)/dr$—Fig. 7.4(c)—and $\partial g(r)/\partial\rho$—Fig. 7.4(h)—may combine to produce a large negative product which will make the evaluation of equation (7.11b) straightforward.

There is a sense in which the numerical demonstration of equation (7.10) is the crucial test for a theory of the equilibrium properties of a liquid. Many of the difficulties in calculating these properties arise from the behaviour of the liquid over distances comparable to the interatomic spacing, and in particular to the position of the minimum in $u(r)$. Because the integral in equation (7.10) is sensitive to these regions, it is a good test of a theory. Clearly the values of $g(r)$ and $u(r)$ used in this integral must be self consistent. For example there would be no point in using the perfect-gas approximation for $g(r)$ unless the value of $g(r)$, calculated from the chosen $u(r)$ by the methods of Chapter 5, were approximately of this form. Also there would be no point in using an experimental $g(r)$ together with a theoretical form for $u(r)$ which is inconsistent with it. Although this point is superficially trivial, it is important in practice because of the difficulty of testing (independently) for self-consistency between the chosen $g(r)$ and $u(r)$. To illustrate this problem Table 7.1 shows the values of the left-hand side of equation (7.10) calculated by Johnson et al.[24] from measured $g(r)$ values and data on $u(r)$ calculated via the $g(r)$–$u(r)$ equations.

TABLE 7.1

**Values of integral in equation (7.10)
using experimental $g(r)$**

Element	YBG—$u(r)$	PY—$u(r)$
Na	-1.6	-6
Pb	-6	-9
A	-3	-6

$$\text{Correct value } +1.000 \begin{pmatrix} -0.005 \\ +0 \end{pmatrix}$$

The discrepancy between the calculated and true values includes the experimental error in $g(r)$ and the lack of consistency between $g(r)$ and $u(r)$. It is expected that both are important.

A possible approach is to use equation (7.10) to modify the solution of, for example, the YBG equation. In principle the YBG equation will give an expression for $g(r)$ valid at large r, and becoming unsatisfactory as r decreases. In contrast equation (7.10) is sensitive to small r. For a L–J potential (equation 3.10) it is shown readily that equation (7.10) reduces to—

$$\frac{16\pi\rho \in \sigma^2}{kT} \int_0^\infty g(r) \left[\left(\frac{\sigma}{r}\right)^4 - 2\left(\frac{\sigma}{r}\right)^{10} \right] dr = 1 \qquad (7.12)$$

This relationship could be used to correct the L–J/YBG calculation of $g(r)$ at low r, in order to obtain a self-consistent combination of $g(r)$ and $u(r)$. Unfortunately this method does not lead to an independent use of equation (7.10), and at the present time this ideal is not practicable.

7.7 Conclusions on Equilibrium Properties

The main subject of the preceding seven Chapters has been the equilibrium properties of liquids. Starting from the pair potential approximation, expressions were obtained for the basic quantities, the equation of state and the specific heat. In order to test these results it would be proper to measure $u(r)$ and $g(r)$ to a high accuracy and insert the experimental data into the relevant expressions. As explained in this Chapter, that procedure requires higher precision than is currently available and thus is not successful. Nevertheless the experimental data do determine quantitatively the main features of these two functions. A theoretical discussion of $u(r)$ and $g(r)$ was given, and it was shown that the theoretical and experimental data were in broad agreement. Even with the addition of the theoretical results to the experimental results, it was not possible to improve the data to the point where the pair theory could be properly tested. In any case, these theoretical results require additional (and serious) approximations beyond the pair approximation, and therefore leave something to be desired. Thus the final result is disappointing, a theory of equilibrium properties is available but cannot be adequately tested. It should be noted that in contrast, for dense gases these difficulties are less severe and a pair theory has had substantial success.[5]

It might be asked, can this theory ever be tested properly? In the case of the specific-heat data for non-conducting liquids, there is no difficulty in principle. Careful experimental work should make data on $(\partial g(r)/\partial T)_V$ available in the course of time. These data when combined with the present knowledge of $u(r)$ should enable absolute values of C_V to be calculated to an accuracy of 10% and a relative accuracy much higher. Even for liquid metals, some progress can be made in this direction, if reliable data for the minimum of $u(r)$ are obtained. The theoretical estimates of $(\partial g(r)/\partial T)_V$ however are likely to be inaccurate, since, as stated earlier, they require some knowledge of both $n^{(3)}$ and $n^{(4)}$ as well as $n^{(2)}$. In this sense an adequate liquid theory is not available, since the higher-order distribution functions are poorly understood. A complete theory should give adequate predictions for these functions. In spite of these difficulties the theoretical estimates of $(\partial g(r)/\partial T)_V$ may approach the accuracy of the experimental data.

In the case of pressure data, the present theory involves a large cancellation of terms, which leads to inaccurate numerical results. It seems unlikely that the basic data, near the triple point, will be improved sufficiently to make a good test possible unless equations like (7.11b) can be exploited. Thus future progress depends upon a theoretical analysis of the reasons for the cancellation of terms, and herein may lie the secret of an adequate pair theory of liquids. If it can be shown what features of condensed matter cause the cancellation, it may be possible to recast the theory in a form that automatically includes this effect. Such a theory would be a basic theory of condensed matter, and hence could form the foundation of a pair theory of the equilibrium state of liquids cast in a form that may be tested easily.

CHAPTER 8

Space and Time Dependent Correlation Functions

At this stage, a further set of concepts will be introduced in order to lay the foundation for a discussion of the transport properties of a liquid. These ideas all involve the motion of the atoms in the liquid, a subject that has not been relevant to the earlier discussion. In this and the following Chapter, the properties of correlation functions that involve time as well as position are discussed. Of greatest importance is the time dependent pair correlation function, $G(r, \tau)$, which is defined in the following Section. Then the methods by which it may be studied experimentally are outlined and compared.

The discussion will be extended at this point to include the significance of quantum effects. Broadly, there are three ways in which quantum effects enter: (i) via the particle statistics; (ii) via position uncertainty; and (iii) via a time uncertainty. Situations involving (i) are not important for most liquids, and their discussion is postponed until Chapter 16. However both position and time uncertainties can be important, particularly the time uncertainty. As an illustration of the importance of quantum effects, this Chapter concludes with a discussion of the phenomenon of detailed balance. Chapter 9 is devoted to a general discussion of the magnitude of quantum corrections to the classical correlation functions, and the method of proceeding via the classical function to the macroscopic limit of the correlation functions. At Chapter 10 the discussion of transport phenomena commences.

8.1. The van Hove Distribution Function

The discussion given in earlier Chapters has concerned the positions of molecules in the liquid. However a complete treatment of the liquid requires a discussion of the position and momentum of each molecule at each instant of time. Thus it is useful to define more general distribution functions involving position (**r**) momentum

$(\mathbf{p})$ and time (t), which describe the distribution at some time t_2 relative to the initial distribution at t_1. The simplest of these is the general pair distribution function, namely $f(\mathbf{r}_1, \mathbf{p}_1, t_1 ; \mathbf{r}_2, \mathbf{p}_2, t_2)$ is the probability of finding an atom at $\mathbf{r}_2$ with momentum $\mathbf{p}_2$ at time t_2 if there was an atom at $(\mathbf{r}_1, \mathbf{p}_1, t_1)$. Fortunately many transport properties may be discussed in terms of a simpler function (G) involving only space and time, i.e.—

$$G(\mathbf{r}_1 - \mathbf{r}_2, t_2 - t_1) = \iint f(\mathbf{r}_1, \mathbf{p}_1, t_1 ; \mathbf{r}_2, \mathbf{p}_2, t_2) \, d\mathbf{p}_1 \, d\mathbf{p}_2 \qquad (8.1)$$

Physically, G is proportional to the probability of finding an atom at $\mathbf{r}_2$ at the time t_2 if there was an atom at $(\mathbf{r}_1, t_1)$. As in the earlier discussion of $g(\mathbf{r}_1 - \mathbf{r}_2)$, it depends only upon the space and time differences because the system is in thermal equilibrium. These differences will be denoted by $\mathbf{r}_1 - \mathbf{r}_2 = \mathbf{r}$ and $t_2 - t_1 = \tau$. $G(\mathbf{r}, \tau)$ is known as the "van Hove correlation function," after its originator.

Figure 8.1 shows how the mathematical definition of $G(\mathbf{r}, \tau)$ is obtained. Figure 8.1(a) shows the location of the ith atom at $\mathbf{r}' - \mathbf{r} = \mathbf{r}_i(0)$ for $\tau = 0$; the mathematical statement of this situation is $\delta[\mathbf{r} + \mathbf{r}_i(0) - \mathbf{r}']$ (integration over the δ function gives unity because only one atom is involved). Figure 8.1(b) shows the location of the jth atom at $\mathbf{r}' = \mathbf{r}_j(\tau)$ for $\tau = \tau$; again the mathematical statement of this situation is $\delta[\mathbf{r}' - \mathbf{r}_j(\tau)]$. Thus the combined situation of Fig. 8.1(a) followed by Fig. 8.1(b) is written—

$$\delta[\mathbf{r} + \mathbf{r}_i(0) - \mathbf{r}'] \cdot \delta[\mathbf{r}' - \mathbf{r}_j(\tau)]$$

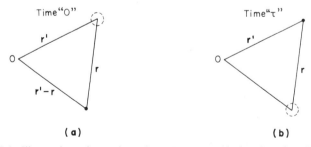

(a)   (b)

FIG. 8.1. Illustration of meaning of van Hove correlation function. The origin is at 0. At time 0—(a)—there is an atom at the point marked by the full circle, and the dotted circle indicates a position where there may or may not be an atom. At time τ—(b)—there is an atom at the point marked by the full circle while there may be or may not be an atom at the dotted circle.

Since the origin can be anywhere in the liquid, this result is integrated over $\mathbf{r}'$, i.e.—

$$\int_V d\mathbf{r}'\delta[\mathbf{r} + \mathbf{r}_i(0) - \mathbf{r}'] \cdot \delta[\mathbf{r}' - \mathbf{r}_j(\tau)]$$

and as the next step all possible pairs (i, j) are taken into account, i.e.—

$$\frac{1}{N}\sum_{ij}\int_V d\mathbf{r}'\delta[\mathbf{r} + \mathbf{r}_i(0) - \mathbf{r}']\delta[\mathbf{r}' - \mathbf{r}_j(\tau)]$$

Finally, a thermal average must be taken, so giving the definition of $G(\mathbf{r}, \tau)$ as—

$$G(\mathbf{r}, \tau) = \frac{1}{N}\left\langle \sum_{ij}\int_V d\mathbf{r}'\delta[\mathbf{r} + \mathbf{r}_i(0) - \mathbf{r}']\delta[\mathbf{r}' - \mathbf{r}_j(\tau)] \right\rangle \qquad (8.2)$$

For a classical system, the vectors $\mathbf{r}_i$ and $\mathbf{r}_j$ commute and equation (8.2) can be rewritten—

$$G(\mathbf{r}, \tau) = \frac{1}{N}\left\langle \sum_{ij}\delta[\mathbf{r} + \mathbf{r}_i(0) - \mathbf{r}_j(\tau)] \right\rangle \qquad (8.3)$$

It is easy to see that if $\tau = 0$, there is a simple connection between $G(\mathbf{r}, 0)$ and $g(r)$ (equation 2.9), i.e.—

$$G(\mathbf{r}, 0) = \delta(r) + \rho g(r) \equiv G_s(\mathbf{r}, 0) + G_d(\mathbf{r}, 0) \qquad (8.4)$$

The sum at equation (8.3) has been divided into two parts; a "self" part corresponding to the terms in which $i = j$ and a "distinct" part involving terms for which $i \neq j$. These two parts are denoted by the subscripts s and d, respectively. Also it should be emphasized that the usual convention (employed in equation 8.4) is to normalize $g(r)$ to unity and $G(\mathbf{r}, \tau)$ to ρ in the limit $\mathbf{r} \to \infty$.

Another convenient form for $G(\mathbf{r}, \tau)$ is obtained by defining a particle-density function $\rho(\mathbf{r}, \tau)$, i.e.—

$$\rho(\mathbf{r}, \tau) = \sum_i \delta[\mathbf{r} - \mathbf{r}_i(\tau)] \qquad (8.5)$$

In terms of this function, van Hove's function may be written—

$$G(\mathbf{r}, \tau) = \frac{\langle \rho(0, 0)\rho(\mathbf{r}, \tau)\rangle}{\rho} \qquad (8.6)$$

which shows that $G(\mathbf{r}, \tau)$ is related to the thermally excited density fluctuations. For sufficiently large r or τ the two atoms (Fig. 8.1) are statistically independent, so that equation (8.6) approaches ρ. The behaviour of $G(\mathbf{r}, \tau)$ for several values of τ is illustrated in Fig. 8.2.

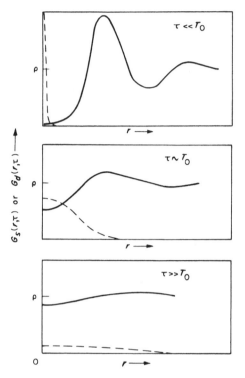

FIG. 8.2. Qualitative behaviour of $G(\mathbf{r}, \tau)$ for several values of τ. The time T_0 is a relaxation time for the liquid structure (from van Hove[27]).

8.2. The Measurement of $G(\mathbf{r}, \tau)$ by Neutron Scattering

(The discussion of neutron scattering given in this Section is abbreviated from Lomer and Low.[26])

In Chapter 6 the cross-section for monoenergetic neutrons scattered through a particular angle was calculated, and related to $g(r)$. However it is possible to measure the energy spectrum of the scattered neutrons and relate the cross-section for a particular scattering angle and energy transfer ($d^2\sigma/d\Omega\, d\omega$) to $G(\mathbf{r}, \tau)$. To do this the "Doppler effect" for the scattering of a wave by a set of moving scattering centres is calculated.

The Fermi pseudo-potential for the ith nucleus—equation (6.12)—is now written as a function of time, i.e.—

$$V(r') = \frac{2\pi\hbar^2}{m_n} \sum_i b_i \delta(\mathbf{r}' - \mathbf{r}_i(t_1)) \qquad (8.7)$$

The scattering amplitude (equation 6.5) is now a function of θ and t_1. However, the scattering process involves a transition of the system between states of energy (say) E' and E''. In terms of the neutron energies before (E_0) and after (E) scattering, it is seen that, $E' - E'' = E_0 - E = \hbar\omega$, and the condition for the conservation of energy may be written as a factor—

$$\delta(E_0 - E - \hbar\omega) = \frac{1}{2\pi} \int e^{-i\omega t} e^{it(E_0 - E)/\hbar} \, dt$$

$$= \frac{1}{2\pi} \int e^{-i\omega t} e^{it(E' - E'')/\hbar} \, dt$$

In formulating[6] the energy of the liquid, a factor $e^{-it(E' - E'')/\hbar}$ appears which cancels the second factor in the above expression. Thus energy conservation is included by writing the scattering amplitude in the form—

$$f(\theta, \omega) = \frac{1}{T} \int_0^T e^{-i\omega t_1} \sum_i b_i \int \delta[r' - r_i(t_1)] e^{iQ.r'} \, dr' dt_1 \quad (8.8)$$

where T is a time long compared with the neutron time of flight. To obtain the scattering cross-section it is necessary to consider the flux in a given energy range ΔE, which is defined by $T\Delta E = 2\pi\hbar$ (or $T\Delta\omega = 2\pi$). Hence through the use of equation (6.13), the cross-section is—

$$\frac{d^2\sigma}{d\Omega \, d\omega} = \frac{1}{2\pi N} \frac{v}{v_0} \int\int\int\int e^{-i\omega(t_1 - t_2)} \left\langle \sum_{ij} b_j^* e^{iQ.(r' - r'')} \delta[r'' - r_j(t_2)] \right.$$

$$\left. \times b_i \delta[r' - r_i(t_1)] \right\rangle \frac{dt_1 \, dt_2}{T} \, dr' \, dr''$$

$$= \frac{1}{2\pi N} \frac{k}{k_0} \int\int e^{i(Q.r - \omega\tau)} \, dr \, d\tau$$

$$\times \left\langle \sum_{ij} \int_V b_i \delta[r + r_j(0) - r'] b_j \delta[r' - r_j(\tau)] \, dr' \right\rangle \quad (8.9)$$

where v_0, v (or k_0, k) are the velocities (or wave numbers) before and after scattering. The factor v/v_0 arises from the conversion of neutron density (i.e. $|f|^2$) to neutron flux.

It can be seen immediately that equation (8.9) is related to the Fourier transformation of equation (8.2). A "scattering law" $S(Q, \omega)$

may be defined* as the transform of $G(\mathbf{r}, \tau)$, i.e.—

$$S(Q, \omega) = \frac{1}{2\pi} \int e^{i(\mathbf{Q}\cdot\mathbf{r} - \omega\tau)}[G(\mathbf{r}, \tau) - \rho] \, d\mathbf{r} \, d\tau \qquad (8.10)$$

and then equation (8.9) can be expressed in terms of $S(Q, \omega)$. To do this, the nuclear scattering lengths must first be taken out of equation (8.9). If there is no correlation between the nuclear spins themselves or between the nuclear spins and the particle positions, the cross-section (equation 8.9) may be expressed in terms of the coherent and incoherent scattering lengths (equations 6.10 and 6.11) through the following relationship—

$$\langle b_i b_j \rangle = b_{coh}^2 + b_{incoh}^2 \delta_{ij}$$

This is valid for a system of Boltzmann particles with a spin independent Hamiltonian and a partition function given by equation (2.12).

In this case, equation (8.9) can be rewritten as—

$$\frac{d^2\sigma}{d\Omega \, d\omega} = \frac{1}{2\pi} \frac{k}{k_0} \int\int e^{i(\mathbf{Q}\cdot\mathbf{r} - \omega\tau)}[b_{coh}^2 G(\mathbf{r}, \tau) + b_{incoh}^2 G_s(\mathbf{r}, \tau)] \, d\mathbf{r} \, d\tau \qquad (8.11)$$

$$= \frac{k}{k_0}[b_{coh}^2 S(Q, \omega) + b_{incoh}^2 S_s(Q, \omega)] + \delta(Q) \text{ terms} \qquad (8.12)$$

where the "self" correlation function is given by—

$$G_s(\mathbf{r}, \tau) = \frac{1}{N} \left\langle \sum_i \int d\mathbf{r}' \delta[\mathbf{r} + \mathbf{r}_i(0) - \mathbf{r}'] \delta[\mathbf{r}' - \mathbf{r}_i(\tau)] \right\rangle \qquad (8.13)$$

For spin-dependent scattering lengths, the cross-section may be separated experimentally into the two components in equation (8.12) by recording those neutrons that are depolarized upon scattering. The "depolarized" cross-section is[26]—

$$\left(\frac{d^2\sigma}{d\Omega \, d\omega}\right)_d = \frac{2}{3} \frac{k}{k_0} b_{incoh}^2 S_s(Q, \omega) \qquad (8.14)$$

and hence enables $S_s(Q, \omega)$ to be measured separately.

8.3. The Measurement of $G(\mathbf{r}, \tau)$ by Scattering of Electromagnetic Radiation

The Hamiltonian[6,7] for the electromagnetic wave (treated by classical electromagnetic theory) and the scattering system is obtained

---

* A constant term ρ is subtracted from G so that a $\delta(Q)$ term is not included in $S(Q, \omega)$.

by noting that the electron momentum occurring in the term $\mathbf{p}^2/2m_e$ is increased from $\mathbf{p}_l$ to $\mathbf{p}_l + (e/c)\mathbf{A}(\mathbf{r}_l)$, where $\mathbf{A}(\mathbf{r}_l)$ is the vector potential of the radiation field at the position of the lth electron. Thus the perturbation term (H') in the complete Hamiltonian is[20,28]—

$$H' = \frac{e}{m_e c}\sum_l \mathbf{p}_l \cdot \mathbf{A}(\mathbf{r}_l) + \frac{e^2}{2m_e c^2}\sum_l \mathbf{A}(\mathbf{r}_l) \cdot \mathbf{A}(\mathbf{r}_l)$$

$$= -\frac{1}{c}\int \mathbf{j}_e(\mathbf{r}) \cdot \mathbf{A}(\mathbf{r})\,d\mathbf{r} - \frac{e^2}{2m_e c^2}\int \rho_e(\mathbf{r})\mathbf{A}(\mathbf{r})\mathbf{A}(\mathbf{r})\,d\mathbf{r} \quad (8.15a)$$

where the electron-current density is given by—

$$\mathbf{j}_e(\mathbf{r}) = -\frac{e}{m_e}\sum_l \delta(\mathbf{r} - \mathbf{r}_l)\mathbf{p}_l \quad (8.15b)$$

and the electron-charge density, ρ_e, is given by equation (6.18b). For each term in equation (8.15), there is actually another term due to the interaction with the nucleus, but this term is small. It may be included in the appropriate electron term without loss of generality.

It will be shown later that the small energy transfers occurring in the scattering process can only be observed in the scattering of light, and not, for example, in the scattering of X-rays. It is easy to show that for light the ratio $eA/cp \sim 10^{-4}$, so that the second term in equation (8.15) is negligible compared to the first (this indirectly justifies the use of the semi classical treatment of the Hamiltonian). Since the first term in equation (8.15) contributes[6,7] zero in first-order perturbation theory, it is necessary to use second-order theory in the calculation of the cross-section. In this treatment, the cross-section arises from a virtual process in which the atom is excited from a state v to a state v' and then de-excited from v' to v. The difference between light scattering and X-ray scattering arises only from the fact that for light the energy is small compared to $E_v - E_{v'}$, whereas for X-rays the energy is large compared to many values of $E_v - E_{v'}$. Thus the electrons may be considered as bound particles for light scattering and free particles for X-ray scattering.

Blume[28] discusses this method of calculating the cross-section and shows that for the assumptions used in Section 6.4, i.e. (i) the electrons move rigidly with the nucleus, (ii) the electronic states are not changed in the scattering process, (iii) the energy transfer to the liquid is small compared to the photon energy and (iv) the liquid atoms are identical, the cross-section is given by—

$$\left(\frac{d^2\sigma}{d\Omega\,d\omega}\right)_{light} = \frac{\mathscr{E}^2}{\hbar^2 c^4}\frac{k}{k_0}S(Q, \omega) \quad (8.16a)$$

where—

$$\mathscr{E}^2 = \sum_v \frac{e^{-\hbar v/kT}}{\sum_v e^{-\hbar v/kT}} \sum_{v'} \left| \frac{\langle v|V^+|v'\rangle\langle v'|V^-|v\rangle}{E_v - E_{v'} + \hbar\omega_k} \right|^2 \qquad (8.16b)$$

plus a spontaneous emission term which is negligible because of the large energy transfer involved. V^+ and V^- are the interaction potentials for electrons in one atom for absorption and emission of a photon.

The quantity $\mathscr{E}$ in equation (8.16) is related to a frequency and wave-number dependent dielectric constant. For light scattering, only very small values of the wave number (i.e. macroscopic effects) are significant, and in this limit Landau and Lifshitz[29] show that—

$$\left(\frac{d^2\sigma}{d\Omega\,d\omega} \right)_{\text{light}} = b_L^2 S(Q, \omega) \qquad (8.17a)$$

where—

$$b_L = \frac{2}{3} \frac{k_0^2}{4\pi} \left(\frac{\partial\varepsilon(\omega_0)}{\partial\rho} \right)_T \qquad (8.17b)$$

and k_0, ω_0 are the wave number and frequency of the incident light. This formula applies when both the incident and scattered light are polarized with the electric vector at right angles to the $(\mathbf{k}_0, \mathbf{k})$ plane, ($\mathbf{k}$ is the wave vector of the scattered light). For directions of polarization other than this, an angular factor will enter equation (8.17). The above formulation has implied that the dielectric constant is not temperature dependent, so that $(\partial\varepsilon/\partial T)_\rho \sim 0$. In general, this term is not zero but will contribute weakly to the observed cross-section; it is proportional to the mean-square temperature fluctuations rather than to $S(Q, \omega)$. For the low Q limit being considered here, it contributes intensity at the same frequencies as the term in equation (8.17), and so produces a slight distortion of the observed spectral intensity from the actual $S(Q, \omega)$. In addition to these terms, there is another term in the complete light scattering formula that is related to the transverse and rotational modes of motion of the system. This term will be discussed in Chapter 13. The latter term may be separated, since [29] the former processes do not alter the polarization of the light, whereas the latter process leads to a rotation of the plane of polarization.

In contrast to the case of neutron scattering, the scattering of electromagnetic radiation is coherent (amplitude and phase of the scattered wavelets from different atoms are the same), and consequently the function $S_s(Q, \omega)$ cannot be measured by this method.

8.4. Comparison of Several Radiation Scattering Techniques

The usefulness of a given technique in measuring the scattering law $S(Q, \omega)$ depends upon its wavelength and energy and the ease with which these can be measured. Table 8.1 summarizes some data concerning neutrons, electrons and several kinds of electromagnetic radiation.

TABLE 8.1

Wavelength and energy of various types of radiation

Radiation	Wavelength (Å)	Approximate momentum $= 2\pi/\lambda$ Å^{-1}	Energy	Approximate δE (expt) (eV)
Neutrons	1–10	2	100–1 meV	10^{-1}–10^{-3}
Electrons	~0·1	0·1	~50 keV	Wide
X-rays	0·5–2	1	~10 keV	Wide
γ-rays	1–2	1	~10 keV	~10^{-9}
Light (Raman)	4000	10^{-3}	~10 eV	~10^{-1}
Light (Brillouin)	4000	10^{-3}	~10 eV	~10^{-5}

δE(expt) indicates the energy-transfer range that may be covered.

The experimental arrangement is that shown in Fig. 6.1, except that the wavelength of the scattered radiation is measured by a spectrometer placed between the sample and the detector.

Since 2π divided by the spacing of atoms in a monatomic liquid is several inverse Ångstrom units and the potential well depth is 10^{-1} to 10^{-2} eV, the ideal choice of momentum and energy transfers (columns 3 to 5 in the Table) would be of this order. Examination of the Table shows that one or other requirement is satisfied for each case listed, but neutrons are the only radiation for which both are satisfied. For electrons, high energies have to be used to satisfy the Born approximation, and then the wavelength is small. For this reason, low-angle scattering is used to measure $d\sigma/d\omega$ and energy analysis to the required accuracy is not possible. For X-rays, momentum analysis is straightforward, but energy analysis is impossible at present. In contrast, Mössbauer γ-rays provide a similar momentum to X-rays, but the energy resolution is too high for the analysis required here. In the case of light, the momenta are very small, but the energy resolution is in the correct range.

The situation depicted in Table 8.1 is further illustrated in Fig. 8.3, where the momentum and energy transfer ranges accessible to the experimentalist are marked in (ω, Q) space. A logarithmic scale has been used so that the axes may be placed at the desired values of

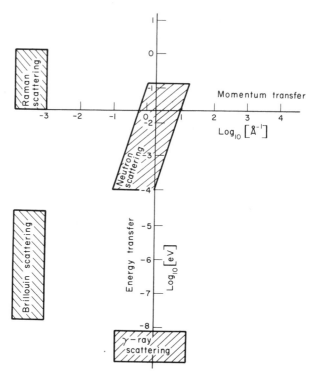

Fig. 8.3. Regions in momentum–energy transfer space covered by the radiation techniques listed in Table 8.1. (From Egelstaff, "Use of Pulsed Neutron Sources in Solid and Liquid State Physics," Fig. 1: presented to IAEA Meeting on "Research Applications of Repetitively Pulsed Boosters," Dubna, 1966.)

energy and momentum. Neutron data may be taken near the intersection of the axes, and other techniques provide data in extreme regions. It is notable that there are large regions of (ω, Q) space that are not accessible at the present time.

The techniques for $G(\mathbf{r}, \tau)$ measurements are neutron and light scattering, and in terms of the wave vectors of the radiation before $(\mathbf{k}_0)$ and after $(\mathbf{k})$ scattering the momentum and energy transfers are—

Neutrons

or

and

$$\left.\begin{aligned}
\mathbf{Q} &= \mathbf{k}_0 - \mathbf{k} \\[2mm]
\frac{\hbar^2 \mathbf{Q}^2}{2m_n} &= E_0 + E - 2(EE_0)^{\frac{1}{2}} \cos \theta \\[2mm]
\frac{\hbar^2}{2m_n}(k_0^2 - k^2) &= \hbar\omega = E_0 - E
\end{aligned}\right\} \qquad (8.18)$$

Light

or

and

$$\mathbf{Q} = \mathbf{k}_0 - \mathbf{k}$$

$$Q^2 = k_0^2 + k^2 - 2kk_0 \cos\theta$$

$$c(k_0 - k) = \omega$$

$$(8.19)$$

If the incident wave vector and angle of scatter (θ) are fixed, the values of Q and ω both vary over the spectrum of scattered radiation. This behaviour is illustrated in Fig. 8.4(a) for neutrons and 8.4(b) for light.

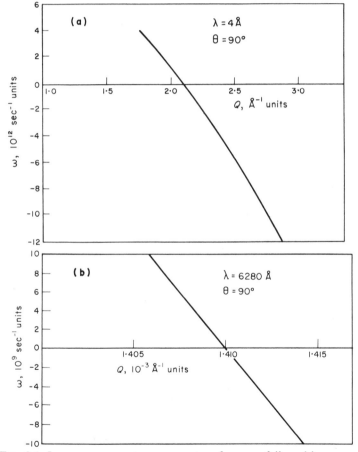

FIG. 8.4. Locus on momentum–energy transfer space followed by a constant angle of scatter experiment; (a), neutron scattering; (b), light scattering.

For neutrons the incident energy may be less than the energy of a transition in the target liquid, and consequently such a transition will not be observed for processes in which the neutron loses energy. This is illustrated by the "cut-off" in Fig. 8.4(a). An effect of this kind does not normally occur in light scattering, where the same transitions can be seen in both energy gain and loss processes. For a given energy transfer $\hbar\omega$ the change (ΔQ) in Q may be evaluated from equations (8.18) and (8.19) as—

$$\left.\begin{aligned} \frac{\Delta Q}{Q} &= -\frac{\hbar\omega}{2E_0} \quad \text{(neutrons)} \\[2mm] &= -\frac{\omega}{ck_0} \quad \text{(light)} \end{aligned}\right\} \tag{8.20}$$

Reference to Table 8.1 shows that the relative change in Q is much greater for neutrons than light, and this is observed in Fig. 8.4.

8.5. Some Properties of $S(Q, \omega)$

$S(\mathbf{Q}, \omega)$ for a liquid is dependent upon the magnitude of $|\mathbf{Q}|$ only, for the same reason as discussed for $S(Q)$—Section 6.5, also the directional averaging inherent in this property shows $S(Q, \omega)$ to be an even function of Q. The connection between $S(Q, \omega)$ and the function $S(Q)$ defined at equation (6.22) is easily demonstrated via the relation (8.4). To do this, the Fourier transformation representation of a δ function is used, i.e.—

$$\frac{1}{2\pi} \int e^{-i\omega t}\, d\omega = \delta(t) \tag{8.21}$$

Thus—

$$\int S(Q, \omega)\, d\omega = \frac{1}{2\pi} \iiint e^{-i\omega\tau}\, d\omega\, d\tau [G(\mathbf{r}, \tau) - \rho]\, e^{i\mathbf{Q}\cdot\mathbf{r}}\, d\mathbf{r}$$

$$= \iint \delta(\tau)[G(\mathbf{r}, \tau) - \rho]\, d\tau\, e^{i\mathbf{Q}\cdot\mathbf{r}}\, d\mathbf{r}$$

$$= \int [G(\mathbf{r}, 0) - \rho]\, e^{i\mathbf{Q}\cdot\mathbf{r}}\, d\mathbf{r}$$

$$= 1 + \rho \int_V e^{i\mathbf{Q}\cdot\mathbf{r}}[g(r) - 1]\, d\mathbf{r} = S(Q) \tag{8.22}$$

Consequently the X-ray or neutron scattering measurements of

$S(Q)$ must involve an integration of $S(Q, \omega)$ over ω at constant Q. In view of the relations (8.18) and (8.20) for neutrons, the intensity as a function of angle does not give a proper integration unless the neutron energy is high compared to $\hbar\omega$ for significant intensity in $S(Q, \omega)$. In this limit, the only correction necessary is that from the laboratory system to the centre of mass system.[30]

At this point, the influence of quantum mechanical effects will be examined, and in order to illustrate their importance, this Chapter is concluded with some remarks on the detailed balance effect in radiation scattering cross-sections. The significance of this effect is related to the fact that the van Hove function $G(\mathbf{r}, \tau)$ is a complex function of time (Section 9.2), and therefore its Fourier transformation will be an unsymmetrical function of ω. This lack of symmetry can be seen to be due to the detailed balancing of the scattering cross-sections. As an example, consider how the radiation would come into thermal equilibrium with a large volume of non-absorbing liquid. Clearly if $S(Q, \omega)$ were symmetrical in ω, equilibrium would never be reached; however, in practice, the assymmetry biases the energy gain or loss processes in a suitable direction until equilibrium is reached, at which point this effect maintains a balance between the two processes. The mathematical form of the asymmetry of $S(Q, \omega)$ can be obtained in the following way.

Consider the ratio of the cross-sections for the scattering of radiation of wave numbers $k_0 \rightarrow k$ to that for wave numbers $k \rightarrow k_0$. The condition of detailed balance requires that the ratio of cross-sections for scattering between two states is equal to the ratio of the statistical weights of the two states, or—

$$\frac{\sigma(k_0 \rightarrow k)}{k^2} \cdot \frac{k_0^2}{\sigma(k \rightarrow k_0)} = \frac{z(\omega_2)}{z(\omega_1)} \tag{8.23}$$

where $z(\omega)$ is the statistical weight of the state of energy ω; for the experiment $k_0 \rightarrow k$, the system is changed from a level $\hbar\omega_1$ to a level $\hbar\omega_2$ (vice versa for $k \rightarrow k_0$). The statistical weight ratio is given by the Boltzmann factor, i.e.—

$$z(\omega_2) = z(\omega_1)\exp\left[\frac{\hbar(\omega_2 - \omega_1)}{kT}\right]$$

$$= z(\omega_1)\exp\left(-\frac{\hbar\omega}{kT}\right) \tag{8.24}$$

since for conservation of energy the energy transfer $\hbar\omega$ is equal to the energy difference of the two states. Finally the cross-section can be

...serted in equation (8.23) from either equation (8.12) or (8.17), to give the result—

$$S(Q, -\omega) = S(Q, +\omega)\, e^{-\hbar\omega/kT} \tag{8.25}$$

It is sometimes convenient to define a symmetrical function of ω, $\tilde{S}(Q, \omega)$ based on equation (8.25), i.e.—

$$\left. \begin{array}{l} S(Q, \omega) = e^{\hbar\omega/2kT}\tilde{S}(Q, \omega) \\[2mm] S(Q, -\omega) = e^{-\hbar\omega/2kT}\tilde{S}(Q, \omega) \end{array} \right\} \tag{8.26}$$

Now using the well known relation connecting the odd (even) part of a function with the imaginary (real) part of its Fourier transformation, it can be shown that—

$$\left. \begin{array}{l} \mathscr{R}\{G(\mathbf{r}, \tau)\} - \rho = \dfrac{1}{(2\pi)^3} \displaystyle\int\!\!\int e^{-(i\mathbf{Q}.\mathbf{r}-\omega\tau)} \left\{ \dfrac{S(Q, \omega) + S(Q, -\omega)}{2} \right\} d\mathbf{Q}.\,d\omega \\[4mm] \mathscr{I}\{G(\mathbf{r}, \tau)\} = \dfrac{1}{(2\pi)^3} \displaystyle\int\!\!\int e^{-i(\mathbf{Q}.\mathbf{r}-\omega\tau)} \left\{ \dfrac{S(Q, \omega) - S(Q, -\omega)}{2i} \right\} d\mathbf{Q}.\,d\omega \end{array} \right\} \tag{8.27}$$

But from equation (8.26)—

$$\frac{S(Q, \omega) - S(Q, -\omega)}{2} = \tanh\!\left(\frac{\hbar\omega}{2kT}\right)\frac{S(Q, \omega) + S(Q, -\omega)}{2} \tag{8.28}$$

And combining equation (8.27) with equation (8.28), a relation is obtained[31] between the real and imaginary parts of $G(\mathbf{r}, \tau)$, i.e.—

$$\mathscr{I}\{G(\mathbf{r}, \tau)\} = -\tanh\!\left(\frac{\hbar}{2kT}\cdot\frac{\partial}{\partial\tau}\right).\,\mathscr{R}\{G(\mathbf{r}, \tau)\} \tag{8.29}$$

where tanh (...) implies an infinite sum of time differentiations obtained by expanding tanh in a power series. Thus to the first order in $\hbar$ equation (8.29) becomes—

$$\frac{2kT}{\hbar}\mathscr{I}\{G(\mathbf{r}, \tau)\} = -\frac{\partial}{\partial\tau}\,\mathscr{R}\{G(\mathbf{r}, \tau)\} \tag{8.30}$$

which will be useful in later Sections.

Thus the Boltzmann statistical factor has been employed to obtain a relationship between the real and imaginary parts of $G(\mathbf{r}, \tau)$, that is to derive an essentially quantum mechanical result. This demonstrates the sensitivity of the time dependence of the correlation functions [or frequency dependence of $S(Q, \omega)$] to quantum mechanical terms. In the next Chapter this question will be explored more fully.

Symbols for Chapter 8

$\mathbf{A}(r)$	Vector potential of radiation field	p_i	Momentum of ith atom
E	Energy state of atom	$z(\omega)$	Statistical weight of state of energy $\hbar\omega$
E' or E''	Energy state of liquid		
$f(\mathbf{p}, \mathbf{r}, \tau)$	Probability function for finding an atom with momentum $\mathbf{p}$, position $\mathbf{r}$, at time τ	$\mathscr{E}$	Scattering amplitude for light scattering by a single atom
		$\varepsilon(\omega)$	Frequency dependant dielectric constant
$\mathbf{j}_e(r)$	Electron current density		

CHAPTER 9

The Classical Limit of $S(Q, \omega)$ and its Relation to Macroscopic Properties

This Chapter is devoted to some general questions that affect the understanding of the liquid state. It begins with a quantitative examination of the classical limit and an account of methods by which first-order quantum corrections may be made to classical results. This is illustrated by a calculation of $S(Q, \omega)$ for the perfect gas: an example that is of practical importance, since it covers one asymptotic limit of the correlation functions. Another problem of importance is the relation between macro- and microscopic properties, and therefore the way of obtaining macroscopic properties from the microscopic correlation functions will be discussed. Part of the discussion of transport properties (Chapter 10 onwards) will be concerned with the macroscopic limiting behaviour of the liquid, and, as a prelude, this Chapter concludes with a discussion of the size of the region in (ω, Q) space to which macroscopic properties may be applied.

9.1. The Classical Limit

Although in the discussion of radiation scattering theory and pair potential theory it has been necessary to appeal to quantum mechanics, the theory of liquids itself has been discussed in the classical limit and the justification of this assumption will be considered now. The classical limit is obtained by allowing $\hbar$ to approach zero, and the region over which such an approximation is valid can be found most easily by examining the function $S(Q, \omega)$. Clearly both the distance and time scales must be large if a classical discussion is to be valid and this requires—

(i) the motions are observed on a time scale $\gg \hbar/kT$, i.e. $1/\omega \gg \hbar/kT$,

(ii) the distances moved are large compared to the atomic wavelength, or for a free atom $2\pi/Q \gg h/(2MkT)^{\frac{1}{2}}$.

To test condition (i) the maximum value of ω (written $\omega_{\max}$) must be found. This will be maximum frequency of vibration of the atom in the field produced by its neighbours, that is the equivalent of the "Debye temperature" for the system. For the present purpose, the maximum frequency (Chapter 14) measured in neutron scattering experiments will be equated with ω_{max}. To test condition (ii) the effective temperature the atom is required and this is assumed to be of the order of the potential well depth, i.e. $kT_{eff} \sim \varepsilon$. Table 9.1 summarizes the results of this comparison. Column 3 shows that for the liquid metals, the first test is well satisfied, but for the rare-gas liquids, argon and neon, this test is not completely satisfactory. Column 4 is closely related to the parameter $1/\Lambda^* = \sigma/2\pi \, (\sqrt{2M\varepsilon}/\hbar)$, introduced by Uhlenbeck and Beth[32] to discuss quantum effects in fluids. The value of Q should be small compared to Q_{max} for the classical limit to be taken. Column 5 shows the value of Q for the principal peak of $S(Q)$—e.g. the main peak in Fig. 6.2. This test is satisfied rather better than the former one, although for neon it is not very satisfactory. In the case of helium, neither the ω test nor the Q test is passed, rather the opposite of limits (i) and (ii) above holds, so that liquid helium is called a quantum liquid.

It should be noted that at high values of Q and ω, outside the limits given in Table 9.1, the function $S(Q, \omega)$ will show quantum effects in every case. However if a liquid is "classical" in the sense defined above, the magnitude of S will be small at these values of Q and ω.

TABLE 9.1

Test of classical limits

Liquid (1)	Temperature (°K) (2)	$\dfrac{\hbar\omega_{max}}{kT}$ (3)	$Q_{max} = \dfrac{\sqrt{2M\varepsilon}}{\hbar}$ (A^{-1}) (4)	Q at peak of $S(Q)$ (A^{-1}) (5)
Helium	2·2	6	0·8	1·9
Neon	26	1	4·5	2·3
Argon	86	1	15	2·0
Sodium	390	0·25	15	2·0
Lead	620	0·1	70	2·2

9.2. The First-order Quantum Correction

The above discussion of the classical limit has shown that a small correction to a classical model may be needed in some cases. Fortunately, the first-order correction may be calculated from equation

(8.30). As a first step the Hermitian symmetry of $G(\mathbf{r}, \tau)$ is demonstrated as follows[27]—

$$G(\mathbf{r}, \tau) = \frac{1}{N} \left\langle \sum \int d\mathbf{r}' \delta[\mathbf{r} + \mathbf{r}_i(0) - \mathbf{r}'] \delta[\mathbf{r}' - \mathbf{r}_j(\tau)] \right\rangle$$

Thus—

$$G^*(\mathbf{r}, \tau) = \frac{1}{N} \left\langle \sum \int d\mathbf{r}' \delta[\mathbf{r}' - \mathbf{r}_j(\tau)] \delta[\mathbf{r} + \mathbf{r}_i(0) - \mathbf{r}'] \right\rangle$$

on changing variables to $\mathbf{r}'' = \mathbf{r}' - \mathbf{r}$—

$$= \frac{1}{N} \left\langle \sum \int d\mathbf{r}'' \delta[\mathbf{r} - \mathbf{r}_j(\tau) + \mathbf{r}''] \delta[\mathbf{r}_i(0) - \mathbf{r}''] \right\rangle$$

on using the stationarity property of correlation functions—

$$= \frac{1}{N} \left\langle \sum \int d\mathbf{r}'' \delta[\mathbf{r} - \mathbf{r}_j(0) + \mathbf{r}''] \delta[\mathbf{r}_i(-\tau) - \mathbf{r}''] \right\rangle$$

on using the even character of δ functions

$$= G(-\mathbf{r}, -\tau) \tag{9.1}$$

Owing to the symmetry of the liquid $G(\mathbf{r}, \tau)$ is an even function of $|\mathbf{r}|$ and the above equation then shows that $\mathscr{I}\{G\}$ is an odd function of τ, whereas $\mathscr{R}\{G\}$ is an even function of τ.

At this stage, it is useful to consider the "intermediate" scattering function $I(Q, \tau)$ defined as—

$$I(Q, \tau) = \int_V e^{i\mathbf{Q} \cdot \mathbf{r}} [G(\mathbf{r}, \tau) - \rho] \, d\mathbf{r} \tag{9.2}$$

For a classical system, I is a real even function of τ and may be expanded as a power series in τ^2 for a certain range of τ—

$$I_{Cl}(Q, \tau) = \sum_{n=0} a_n(Q) \tau^{2n} \tag{9.3}$$

where $a_n(Q)$ are coefficients defined by equation (9.3) and the subscript "Cl" implies classical. Equation (8.30) may be rewritten (by using equation 9.2) as—

$$\frac{2kT}{\hbar} \mathscr{I}\{I(Q, \tau)\} = -\frac{\partial}{\partial \tau} \mathscr{R}\{I(Q, \tau)\} \tag{9.4}$$

and then I_{Cl} may be inserted in the right-hand side of this equation.

In this way, the first-order correction is found to be—

$$\mathcal{I}\{I(Q, \tau)\} \simeq -\frac{\hbar}{2kT} \sum_{n=1} 2n a_n(Q) \tau^{2n-1} \tag{9.5}$$

It is easy to see that equation (9.5) can be obtained from equation (9.3) via the substitution—

$$\tau \equiv \tau - i\frac{\hbar}{2kT} \tag{9.6}$$

suggested by Schofield[31] or via the substitution—

$$\tau^2 \equiv \tau^2 - i\frac{\hbar\tau}{kT} \tag{9.7}$$

suggested by Egelstaff.[33] Thus the conversion of a classical correlation function to a new function correct to first order in $\hbar$ may be accomplished by either of the time substitutions (9.6) or (9.7). The former has the advantage that it can be applied directly to $S(Q, \omega)$ by using equation (8.26)—$\tilde{S}$ is equivalent to S_{Cl}—whereas the latter has the advantage that it also gives the correct expression for some of the $\hbar^2$ terms. The method described above is not a formal proof, since it has not been shown that the substitutions (9.6) and (9.7) give the correct form at large τ. Although this may be done, it should be noted that the large τ region is relatively unimportant in practice, since there $\tau \gg \hbar/kT$.

9.3. The Classical Limit of $S(Q, \omega)$

It was pointed out in Section 8.5 that the classical limit of $S(Q, \omega)$ yields a function that is symmetrical in ω. This is equivalent to the statement—

$$\omega \gg \bar{\omega}(Q) \tag{9.8}$$

where $\bar{\omega}(Q)$ is the mean value of ω at a given value of Q. Equation (9.8) is the most useful condition for specifying this limit, since it can be shown that $\bar{\omega}$ is independent of the dynamics of the scattering system.

The equation connecting the scattering law and the intermediate scattering function is—

$$I(\mathbf{Q}, \tau) = \int_{-\infty}^{\infty} S(Q, \omega)\, e^{i\omega\tau}\, d\omega$$

$$= \int_{-\infty}^{\infty} S(Q, \omega)\, d\omega + i\tau \int_{-\infty}^{\infty} \omega S(Q, \omega)\, d\omega$$

$$- \frac{\tau^2}{2} \int_{-\infty}^{\infty} \omega^2 S(Q, \omega)\, d\omega + \text{etc.} \tag{9.9}$$

Thus the mean value of ω is related to the term linear in τ in $I(Q, \tau)$, and this is clearly the first term in the expansion of $\mathscr{I}\{I\}$. An assumption, valid for all cases in which the interatomic forces are velocity independent, is that the term linear in τ is linear in $\hbar$ also. This assumption covers the force laws discussed in Chapters 3 and 4. In this case, the mean value of ω can be calculated from equation (9.4), and the classical value of $\overline{\omega^2}$. Thus—

$$\frac{2kT}{\hbar}[\tau\bar{\omega} + O(\tau^3)] = -\frac{\partial}{\partial\tau}\left[1 - \frac{\tau^2}{2}\overline{\omega_{Cl}^2} + O(\tau^4)\right]$$

or—

$$\bar{\omega} = \frac{\hbar}{2kT}\overline{\omega_{Cl}^2} \tag{9.10}$$

where—

$$\bar{\omega} = \frac{1}{S(Q)}\int \omega S(Q, \omega)\,d\omega$$

and—

$$\overline{\omega_{Cl}^2} = \frac{1}{S(Q)}\int \omega^2 S_{Cl}(Q, \omega)\,d\omega$$

It will be shown in Chapter 11 (equation 11.24b) that—

$$\overline{\omega_{Cl}^2} = \frac{kT}{M}\frac{Q^2}{S(Q)} \tag{9.11}$$

thus from equation (9.10)—

$$\bar{\omega} = \frac{\hbar Q^2}{2MS(Q)} \tag{9.12}$$

and from equation (9.8)—

$$\omega \gg \frac{\hbar Q^2}{2MS(Q)} \tag{9.13}$$

in a classical system.

The values of ω to be used in equation (9.13) are those for which $S_{Cl}(Q, \omega)$ has a significant amplitude. Equation (9.11) gives a rough estimate of the order of magnitude of ω to use. With this substitution the condition required is—

$$\frac{\hbar Q}{2\sqrt{MkTS(Q)}} \ll 1 \tag{9.14}$$

The Q values for which the left-hand side of this equation is approximately 0.5 at the triple point, are 10, 30, 50 and 180 A^{-1} for neon, argon, sodium and lead, respectively. Thus equation (9.14) is satisfied rather better than the conditions tested in Table 9.1. This shows that the step from equation (9.13) to equation (9.14) was slightly optimistic, and that it is preferable to use experimental values of ω in equation (9.13).

It should be noted that the classical limit of $S(Q)$ [$S(Q)$ was defined by equation 6.22] is specified by the same expressions (i.e. equations 9.13 and 9.14), and consequently the remarks made in this Section apply also to the discussion of Chapters 5 to 7. Thus equation (9.13) gives an experimental definition of the classical region of the liquid state, and Table 9.1 shows that most liquids will fall broadly within this region. However small corrections are needed at high ω and Q values, and may be applied through the substitutions at equation (9.6) or (9.7). An example illustrative of these points will be given in the next Section. Apart from this example (and the discussion of quantum liquids in Chapter 16) a classical treatment of the liquid state will continue to be given here.

9.4. The Perfect Gas

In order to calculate either $G(\mathbf{r}, \tau)$ or $S(Q, \omega)$ it is necessary to specify both the statistical behaviour and the kinematical behaviour of the atoms in the system. It has been shown in Section 8.5 that quantum corrections are related to the statistical properties of the medium, and in this Section the example of the perfect gas will be used to show that they can be related to the kinematical behaviour too.

The first step is to consider only the statistical behaviour of the atoms in the gas and to calculate $G(\mathbf{r}, \tau)$ from the Maxwell–Boltzmann distribution of velocities. If the most probable velocity is $v_0^2 = 2kT/M$, the probability that an atom has a velocity $\mathbf{v}$ is proportional to $\exp(-v^2/v_0^2)$. But the displacement of an atom in time τ through the distance $\mathbf{r}$ corresponds to $\mathbf{v} = \mathbf{r}/\tau$, hence the probability of finding it at $(\mathbf{r}, \tau)$ after it was at $(0, 0)$ is—

$$G(\mathbf{r}, \tau) = \frac{1}{\pi^{\frac{3}{2}}|\mathbf{v}_0\tau|^3} \exp -\frac{r^2}{|\mathbf{v}_0\tau|^2} \tag{9.15}$$

which has been normalized by the condition (9.29). On Fourier-transforming this expression, the classical scattering law for a gas is obtained, i.e.,—

$$S(Q, \omega) = \frac{1}{\sqrt{\pi}Qv_0} \exp -\frac{\omega^2}{(Qv_0)^2} \tag{9.16}$$

It is notable that this expression is symmetrical in ω as expected from Section (9.2); in particular it does not contain the detailed balance factor $(\exp - \hbar\omega/2kT)$.

As a second step, this result will be compared with that obtained from a kinematical and statistical calculation of the scattering law. Figure 9.1 shows the addition of momentum $\hbar\mathbf{Q}$ to an atom which has momentum $\mathbf{p}$. For a perfect gas, the energy of an atom is just the kinetic energy or $p^2/2M$, and thus this process leads to a unique value of the final energy, i.e. $|\mathbf{p} + \hbar\mathbf{Q}|^2/2M$. Now the scattering law is the probability of absorption, by the target, of an energy $\hbar\omega$ if it is given an additional momentum $\hbar\mathbf{Q}$. Thus in this case—

$$S_{\mathbf{p}}(\mathbf{Q}, \omega) = \delta\left(-\frac{p^2}{2M} + \frac{|\mathbf{p} + \hbar\mathbf{Q}|^2}{2M} - \hbar\omega\right) \qquad (9.17)$$

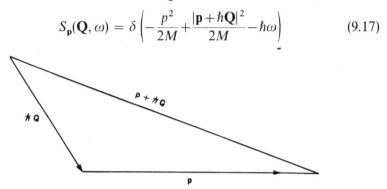

FIG. 9.1. Momentum diagram showing addition of momentum $\hbar\mathbf{Q}$ to a free atom (travelling with momentum $\mathbf{p}$) by radiation scattering.

where the subscript $\mathbf{p}$ denotes that the target atom has momentum $\mathbf{p}$, and the integral over the δ function is unity. The probability of initial momentum (p) is given by the Boltzmann distribution i.e.—

$$f(\mathbf{p}) = \frac{1}{(2\pi MkT)^{\frac{3}{2}}} \exp - \left(\frac{p^2}{2MkT}\right) \qquad (9.18)$$

Thus by combining equations (9.17) and (9.18), both the kinematics and the statistics are included in the discussion, and a comparison of these equations shows that $\hbar$ (and so the quantum effects) occur in the kinematical factor. Moreover, as in the discussion of Section (8.5), the quantum mechanical effects have entered through an apparently classical requirement—namely the recoil of the atom upon being struck by the neutron. This gives the connection between Section 8.5 and this Section, since for a perfect gas the recoil effect gives rise to detailed balancing and the establishment of thermal equilibrium.

Now to find $S(Q, \omega)$, the procedure in equation (8.1) is followed (note $\mathbf{p}_1 \equiv \mathbf{p}$ and $\mathbf{p}_2 \equiv \mathbf{p} + \hbar\mathbf{Q}$), and the product of equations (9.17) and (9.18) is integrated over all values of $\mathbf{p}$ so giving—

$$kTS(\mathbf{Q}, \omega) = kT \int \frac{e^{-p^2/2MkT}}{(2\pi MkT)^{\frac{3}{2}}} \, \delta\left(\frac{\hbar^2 Q^2}{2M} + \frac{\hbar\mathbf{p} \cdot \mathbf{Q}}{M} - \hbar\omega\right) dp$$

$$= -\frac{1}{2\sqrt{\pi\alpha}} \int_\infty^1 \frac{(\alpha + \beta)^2}{4\alpha} \exp - \frac{(\alpha + \beta)^2}{4\alpha\mu^2} \, d\left(\frac{1}{\mu^2}\right)$$

$$= \frac{1}{2\sqrt{\pi\alpha}} \exp - \frac{(\alpha + \beta)^2}{4\alpha} \qquad (9.19)$$

where μ is the cosine of the angle between $\mathbf{p}$ and $\mathbf{Q}$, and—

$$\alpha = \frac{\hbar^2 Q^2}{2MkT}; \qquad \beta = \frac{\hbar\omega}{kT}; \qquad (9.20)$$

The Fourier transform of equation (9.12) gives the van Hove correlation function as—

$$G(r, \tau) = \left[\frac{M}{2\pi\tau(kT\tau - i\hbar)}\right]^{\frac{3}{2}} \exp - \frac{Mr^2}{2\tau(kT\tau - i\hbar)} \qquad (9.21)$$

in agreement with a direct quantum mechanical calculation.[27] This equation reduces to equation (9.15) when $\tau \gg \hbar\omega/kT$ and equation (9.19) reduces to equation (9.16) when $\omega \gg \hbar Q^2/2M$. From the latter condition, it can be seen that the classical limit is reached for either Q small or M large. This sometimes leads to confusion in specifying the classical limit, and it is better to consider these quantities in combination (as in equation 9.13). Equation (8.30) gives the first-order correction to a classical $G(\mathbf{r}, \tau)$; it is a simple matter to insert equation (9.15) in equation (8.30) and confirm that the term in equation (9.21) linear in $\hbar$ is obtained. Also the substitution of equation (9.7) in equation (9.15) gives equation (9.21) immediately.

The importance of these results is that at high Q the scattering law for a liquid reduces to that of a perfect gas, and as seen above the high Q region may involve quantum corrections. This situation arises because high Q values of $S(Q, \omega)$ are related to the short time and small distance region of $G(\mathbf{r}, \tau)$. This region of $G(\mathbf{r}, \tau)$ is dominated by the behaviour of the $\delta(r)$, or self-term of $G(\mathbf{r}, 0)$—equation (8.4). At short times the atom, chosen at the origin, is moving freely and thus its initial motion is like that of a perfect-gas atom. An alternative way of making this point is to note that in order to locate an atom at $(0, 0)$ all the Fourier components of its motion must be specified. This is equivalent to (physically) exciting all these components, hence it behaves initially as a free atom.

9.5. Relation between Macroscopic Properties and $S(Q, \omega)$

Examination of the general definition of $S(Q, \omega)$ (equation 8.10) shows that as $Q \to 0$ an integral over the whole volume is obtained. Thus this limit may be used to obtain macroscopic properties from radiation scattering data and will be so employed many times. Also in this limit the time dependence of $I(Q, \tau)$ becomes negligible, i.e.—

$$[I(Q, \tau)]_{Q \to 0} \to \text{constant} \qquad (9.22)$$

since the number of atoms in the total volume is constant. The Fourier transformation of equation (9.22) is proportional to $\delta(\omega)$, which is the limiting form for $S(Q, \omega)$ as $Q \to 0$. Since $I(0, \tau)$ is independent of τ it is possible to obtain a macroscopic limit from $I(0, 0)$, although at first it may appear unlikely that a $(\tau = 0)$ limit can be related to a macroscopic property. By virtue of equations (8.22) and (6.23), this limit is proportional to the isothermal compressibility. It is useful to obtain the same result directly from fluctuation theory.

The combination of equation (8.4) with equation (2.8) yields—

$$I(0, 0) = \int [G(\mathbf{r}, 0) - \rho]\, d\mathbf{r} = \frac{(\overline{N^2} - \overline{N}) - \overline{N}^2}{\overline{N}} + 1 \qquad (9.23)$$

Now using the result that—

$$\overline{(\Delta N)^2} = \overline{(N - \overline{N})^2} = \overline{N^2} - \overline{N}^2 \qquad (9.24)$$

equation (9.23) can be reduced to—

$$I(0, 0) = \frac{\overline{(\Delta N)^2}}{N} \qquad (9.25)$$

The fluctuation theory described in Chapter 1 yielded a value for $\overline{(\Delta V)^2}|_N$ (equation 1.14), and this is proportional to $\overline{(\Delta N)^2}|_V$ since—

$$\overline{(\Delta V)^2}\bigg|_N \equiv N^2 \overline{\left[\Delta\left(\frac{V}{N}\right)\right]^2} \equiv \frac{V^2}{N^2} \overline{(\Delta N)^2}\bigg|_V \qquad (9.26)$$

Thus from equations (1.14), (9.26) and (9.25)—

$$S(0) = I(0, 0) = kT\frac{N}{V}\chi_T \qquad (9.27)$$

in agreement with equation (6.23). An experimental test of this equation is given in Fig. 9.2, showing that even for helium equation (9.27) is satisfied for $Q \to 0$.

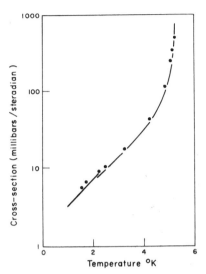

FIG. 9.2. Comparison of measured value of $S(0)$ and value calculated from equation (9.27). The measured cross-sections for liquid helium are shown as solid circles and the calculation as a line. (Egelstaff and London, *Proc. Roy. Soc.*, 1957, **A242**, 374.)

This result should be compared with the corresponding result for the self-correlation function (equation 8.13). In this case the definition of G_s gives—

or—

$$G_s(\mathbf{r}, 0) = \delta(r) \left.\begin{array}{c} \\ \\ \end{array}\right\}$$
$$I_s(\mathbf{Q}, 0) = 1$$

(9.28)

Thus for the self-correlation function, $I_s(Q, 0)$ registers the fact that only one atom is being considered. The same fact means that the integral over the volume (or $Q \to 0$ limit) is unity, i.e.—

$$I_s(0, \tau) = 1 \qquad (9.29)$$

In contrast to the former case, the self-correlation function has no special limiting form for $I_s(0, 0)$. However equation (9.29) shows that when $Q \to 0$, the scattering law $S_s(Q, \omega) \to \delta(\omega)$.

In later Chapters it will be shown how the transport coefficients are related to the $Q \to 0$ limit.

9.6. The Continuum Limit

Both the classical limit (equation 9.13) and the macroscopic co-efficients are obtained at small values of Q. Whereas the classical limit applies at small, but finite, values of Q, it is not clear whether macroscopic concepts can be applied usefully to non-zero values of Q. It is usual to assume that the liquid can be treated as a macroscopic continuum if the microscopic structure of $G(\mathbf{r}, \tau)$ may be neglected. This is the large r and τ region where (equation 8.6)—

$$G(\mathbf{r}, \tau)_{r \to \infty \atop \text{or } \tau \to \infty} \to \rho \qquad (9.30)$$

Since the fluctuations of $G(r, \tau)$ are largest for $\tau = 0$, the continuum region may be defined as the range $r \leq r_c$, where the microstructure of $g(r)$ occurs for $r < r_c$. Thus—

$$G(r \geq r_c, 0) \sim \rho \qquad (9.31)$$

By considering the Fourier transform of $G(r, \tau)$ it is possible to find a useful definition of the region over which $S(Q, \omega)$ is related to the macroscopic properties, namely—

$$S(Q \leq Q_c) \to S(0) \qquad (9.32)$$

where $Q_c \sim 2\pi/r_c$. This region may be found by experiment, or it may be estimated from equation (9.31) and the calculated value of $g(r)$. In most cases Q_c is less than one-tenth of the position of the main diffraction peak (i.e. $\frac{1}{10}$ the value of Q given in column 5 of Table 9.1).

Another definition, based on the idea that the molecules in a liquid are in collision with one another, is sometimes used in hydrodynamic theory. This states that the continuum approximation is applicable if the "mean free path" of the molecules is much smaller than the range of r being observed. In a dense liquid the mean free path is less than the spacing between molecules, and hence this definition is included in the above one. Sometimes this is altered to a definition based on time scales, namely the time scale of an observation should be large compared to the time between collisions or the time of a collision. This too is equivalent to the above definition, since the range of ω over which the continuum theory is useful may be calculated from equation (9.11) by setting $Q \sim Q_c$.

In such ways a continuum region of (ω, Q) space may be defined. It is to this region that the macroscopic parameters apply and in which macroscopic equations of motion may be used. The discussion of transport properties will start with the behaviour in this continuum region and then proceed to an examination of microscopic behaviour.

Symbols for Chapter 9

$a_n(Q)$	Coefficient in the expansion of the intermediate scattering function	α	$\hbar^2 Q^2/2MkT$
v_0	Most probable velocity in a Maxwellian velocity distribution	β	$\hbar\omega/kT$

Diffusion and Single Particle Motion

Chapters 10 to 12 are concerned with one of the basic phenomena occurring in a liquid, namely the allowed motion of a single atom. This Chapter deals with the scale (in distance and time) of the diffusion process, and is thus concerned with those motions that are observed on a relatively long time scale. In contrast, Chapter 11 deals with motion on a relatively short time scale. For this discussion the important quantity is the velocity correlation function, and some methods of measuring or calculating it will be described there. Finally, Chapter 12 outlines a number of phenomenological methods of calculating the diffusion constant itself. The basic difficulty in calculating transport coefficients is described in Section 12.1 and shown diagrammatically in Fig. 12.1. Possibly Fig. 12.1 may help the reader to appreciate the difference between the approach adopted here and that adopted for earlier Chapters.

Chapter 10 commences with a discussion of the random-walk theory of diffusion and continues with accounts, from several points of view, of the basic step in the random walk.

10.1. Einstein's Random-walk Theory

Diffusion rates of $\sim 10^{-5}$ cm²/sec are characteristic of the liquid state (Table 1.1); as compared to diffusion in the solid state, this represents a rapid motion, but on a gaseous picture it is slow. In order to discuss this phenomenon, it will be assumed that diffusion is a cyclic process. That is, starting from a given origin an atom moves (by an unspecified mechanism) a distance l_1 in a time τ_1; at this point its previous history is forgotten and it commences a fresh move of distance l_2 taking a time τ_2, and so on indefinitely. The important idea, which defines the lengths and times involved, is that there is a "loss of memory" between the steps; each step being the minimum one required for loss of memory.

Figure 10.1 illustrates this process. In order to calculate the probability of finding an atom at $(\mathbf{r}, \tau + \tau_1)$ it is assumed that there is an atom at $(\mathbf{r'}, \tau)$ and that it makes a small step $(\mathbf{l}_1, \tau_1)$ uncorrelated with its previous history. The initial vector $\mathbf{r'}$ is regarded as an arbitrary member of the probability distribution $G_s(\mathbf{r'}, \tau)$; that is of the distribution of positions which may be taken up by an atom initially at $(0,0)$. Also since the step $(\mathbf{l}_1, \tau_1)$ is uncorrelated with previous history, the chance of the atom at $(\mathbf{r'}, \tau)$ making this move is $G_s(\mathbf{l}_1, \tau_1)$. Consequently the probability of finding an atom at $(\mathbf{r}, \tau + \tau_1)$ is—

$$G_s(\mathbf{r}, \tau + \tau_1) = \int_V G_s(\mathbf{r} - \mathbf{l}_1, \tau) G_s(\mathbf{l}_1, \tau_1) \, d\mathbf{l}_1 \qquad (10.1)$$

[Note: since the two factors on the right-hand side are uncorrelated, this equation could have been written down with non-thermally averaged functions.]

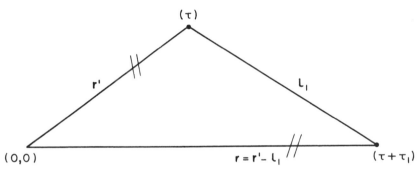

FIG. 10.1. Illustration of random-walk model of diffusion. An atom is initially at $(0, 0)$ and after time τ it is at $(\mathbf{r'}, \tau)$; a short time later it has moved a small distance to $(\mathbf{r'} - \mathbf{l}_1; \tau + \tau_1)$. Full circles denote atomic positions and double bars on $\mathbf{r}$ and $\mathbf{r'}$ indicate that the magnitude of these vectors is much greater than $|\mathbf{l}_1|$.

Now $\mathbf{l}_1$ and τ_1 are small increments compared to $\mathbf{r}$ and τ, so that the left-hand side may be expanded about τ and the right-hand side about $\mathbf{r}$, i.e.—

$$G_s(\mathbf{r}, \tau) + \tau_1 \frac{\partial G_s(r, \tau)}{\partial \tau} + \text{etc.}$$

$$= \int_V G_s(\mathbf{l}_1, \tau_1) \, d\mathbf{l}_1 \left[G_s(\mathbf{r}, \tau) - x \nabla G_s(\mathbf{r}, \tau) + \frac{x^2}{2} \nabla^2 G_s(\mathbf{r}, \tau) + \text{etc.} \right] \qquad (10.2)$$

where x is the component of $\mathbf{l}_1$ along $\mathbf{r}$. Since the integral of G_s over all space is unity, the first terms on both sides of this equation cancel. The second term on the right-hand side is zero, since G_s is even in $\mathbf{l}_1$;

thus on neglecting the higher-order terms, equation (10.2) reduces—

$$\tau_1 \frac{\partial G_s(\mathbf{r}, \tau)}{\partial \tau} \simeq \int_V \frac{x^2}{2} \nabla^2 G_s(\mathbf{r}, \tau) G_s(\mathbf{l}_1, \tau) \, d\mathbf{l}_1$$

$$= \frac{\overline{l_1^2}(\tau_1)}{6} \nabla^2 G_s(\mathbf{r}, \tau) \qquad (10.3)$$

where—

$$\overline{l_1^2}(\tau) = \int_V l_1^2 G_s(\mathbf{r}, \tau) \, d\mathbf{l}_1$$

This is the equation of motion from which $G_s(\mathbf{r}, \tau)$ may be determined (see Section 10.2), and inspection of it shows that the values of τ_1 and l_1 have not been specified, other than to include the loss-of-memory requirement. However it would be convenient if the ratio $\overline{l_1^2}/6\tau_1$ was independent of τ_1, since in that case equation (10.3) would be independent of τ_1 also, and the ratio could be specified by any pair of values of τ_1 and l_1.

To demonstrate this point, it is necessary to show that, for p diffusive steps, the following relationship holds—

$$\langle r^2(p\tau) \rangle = p \langle r^2(\tau) \rangle \qquad (10.4a)$$

where $r(\tau)$ is the distance moved in time τ. The left hand side of this equation may be expanded as follows—

$$\langle r^2(p\tau) \rangle = \left\langle \left| \sum_{i=1}^{p} \mathbf{r}_i(\tau) \right|^2 \right\rangle = \left\langle \sum_{i=1}^{p} r_i^2(\tau) \right\rangle + \left\langle \sum_{i \neq j} \mathbf{r}_i(\tau) \mathbf{r}_j(\tau) \right\rangle$$

on using the assumption that $\mathbf{r}_i$ and $\mathbf{r}_j$ are uncorrelated—

$$\simeq p \langle r^2(\tau) \rangle \qquad (10.4b)$$

Thus the ratio is independent of the number of steps. It is usual to call this ratio the diffusion constant (D) and to specify it by the smallest values of $\overline{l_1^2}$ and τ_1 (denoted by $\overline{l^2}$ and τ_0, respectively) for which the loss-of-memory condition is satisfied, thus—

$$D = \frac{\overline{l^2}}{6\tau_0} = \frac{\overline{l_1^2}}{6\tau_1} \qquad (10.5)$$

where l_1 and τ_1 are any pair of corresponding values. This result demonstrates that equation (10.3) yields the macroscopic limit of $G_s(r, \tau)$, since macroscopic values of l_1 may be used in equation (10.5).

Equations (10.3) with (10.5) will be recognized as the usual Fick's law diffusion equation. A macroscopic concentration gradient is imagined to be set up in the liquid and it is assumed that the flux density is proportional to it. The flux of atoms takes a direction to reduce the concentration gradient, and the constant of proportionality is denoted by D (equation 10.16, for example). Then the fluctuation in the concentration in a volume element $d\mathbf{r}$ at $\mathbf{r}$ is given by $D\nabla^2 G_s$ and is equal to the time rate of change of the concentration, or—

$$D\nabla^2 G_s(\mathbf{r}, \tau) \simeq \frac{\partial G_s(\mathbf{r}, \tau)}{\partial \tau} \tag{10.6}$$

This equation has been derived from the assumption that the liquid behaves as a macroscopic continuum in contrast to equation (10.3), which was derived from a microscopic point of view. It is confirmed, therefore, that the diffusion equation gives the continuum limit to the self correlation function.

10.2. Solution of the Diffusion Equation and Measurement of D

A solution of equation (10.3) or (10.6) is required that has the form $\delta(r)$ as $r \to 0$ and when integrated over the volume is equal to unity. It is simple to check that the following expression is a solution satisfying these requirements—

$$G_s(\mathbf{r}, \tau) = \frac{1}{(4\pi D\tau)^{\frac{3}{2}}} \exp -\frac{r^2}{4D\tau} \tag{10.7}$$

As a check, the mean-square value of r may be calculated, i.e.—

$$\overline{r^2} = \int r^2 G_s(r, \tau)\, d\mathbf{r} = 6D\tau \tag{10.8}$$

in agreement with equation (10.5). Fourier transformation of equation (10.7) is readily accomplished to give—

$$\left.\begin{aligned} I_s(Q, \tau) &= \exp_{\cdot}-Q^2 D\tau \\ S_s(Q, \omega) &= \frac{1}{\pi}\frac{DQ^2}{\omega^2 + (DQ^2)^2} \end{aligned}\right\} \tag{10.9}$$

Figure 10.2 gives the shape of $S_s(Q, \omega)$ in this case. The full width at half height of this function is $2DQ^2$; if $S_s(Q, \omega)$ is measured at a number of fixed values of Q, a graph of width as a function of Q^2 should be a straight line of slope $2D$. This is one test of the correctness of the

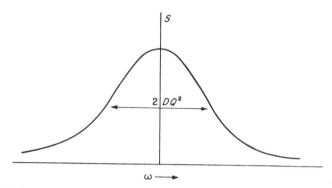

FIG. 10.2. Illustration of shape of $S_s(Q, \omega)$ according to equation (10.9). The curve is a Lorentzian in ω for fixed Q, and the width at half amplitude is a measure of D.

above theory. In addition a departure from this result is an indication that the region of $(\mathbf{r}, \tau)$ space being covered is sufficiently near the origin that the history of the atomic motion is important.

One method of measuring D would be to measure $S_s(Q, \omega)$ and fit to equation (10.9); in this case it is necessary to make measurements at small values of Q and ω to obtain accurate results (Section 9.6). In the case of neutron scattering, the (Q, ω) values are usually too large: in the case of light scattering, the (Q, ω) values are of the correct size, but the function $S_s(Q, \omega)$ cannot be measured as the scattering is coherent. Two other methods are available. One method is to measure $\overline{l_1^2}$ and τ on a macroscopic scale—e.g. by observing the intensity, as a function of position and time, of emission from a radioactive isotope of the element being investigated. In this case, $G_s(\mathbf{r}, \tau)$—equation (10.7)—is being measured directly. Another method of determining D is to measure the diffusion rate for nuclear spins by the nuclear magnetic resonance technique. Essentially, the amplitude of the magnetization is measured as a function of time; for a high diffusion rate, the amplitude decays rapidly. The detailed relation[34] between the amplitude and the diffusion constant depends upon the details of the experimental method. In suitable cases (e.g. molecules containing protons) a high accuracy may be achieved.

It is interesting that for simple liquids all three methods give the same value of D within experimental error, although the neutron method makes measurements over distances of a few Ångstroms, the nuclear magnetic resonance method over distances of several hundred Ångstroms and the tracer method over distances of millimetres. This demonstrates the validity of equation (10.4). In terms of the wave number Q, all of these results are confined to $Q < 0.5 \, \text{Å}^{-1}$

(usually $\ll$). The microscopic structure of the liquid is observed at higher values of Q (Fig. 6.2) and as expected from the discussion of Section 10.1 its effect has not been included.

The next stage in the discussion of diffusion is to obtain the minimum values of l and τ_0. As illustrated by Fig. 12.1, this requires some discussion of the molecular dynamics, and because no general method is available the discussion will be accomplished through the use of models.

10.3. Jump Diffusion

The above theory is satisfactory when $(\mathbf{r}, \tau)$ is large compared to the elementary values of $(\mathbf{l}, \tau_0)$. However, if the experimental values of $(\mathbf{Q}, \omega)$ are such that small values of $(\mathbf{r}, \tau)$ are observed, then it breaks down. In this case an elementary diffusive step can be observed, and the function $S_s(Q, \omega)$ differs significantly from equation (10.9). It is necessary to describe the details of a diffusive step in order to calculate the scattering law.

Assume that an atom remains at a given site for a time τ_0. During this time it may vibrate about a centre of equilibrium building up a small thermal cloud. After the elapse of time τ_0, it is assumed that the atom moves rapidly (i.e. in a time negligible compared to τ_0) to a new site situated at point $\mathbf{l}$ relative to the original site. The length l has to be large compared to the dimensions of the thermal cloud, and the process is repeated indefinitely. Chudley and Elliot[35] have calculated the scattering law for a "jump diffusion" model of this kind. If $G_s(r, \tau)$ is the probability of finding an atom at $\mathbf{r}$, the time rate of change of $G_s(r, \tau)$ is given by a simple rate equation, i.e.—

$$\frac{\partial G_s(\mathbf{r}, \tau)}{\partial \tau} = \frac{\text{fluctuation}}{\text{relaxation time}} = \frac{\frac{1}{n}\sum_{\mathbf{l}}[G_s(\mathbf{r}+l, \tau) - G_s(\mathbf{r}, \tau)]}{\tau_0} \qquad (10.10)$$

where n is the number of sites to which it is possible to jump. In the limit $\mathbf{r} \gg |\mathbf{l}|$, the term $G_s(\mathbf{r}+\mathbf{l}, \tau)$ may be expanded about $\mathbf{r}$ and equation (10.3) is recovered (i.e. r is macroscopic).

Equation (10.10) may be Fourier-transformed to give an equation in the intermediate scattering function, i.e.—

$$\frac{\partial I_s(Q, \tau)}{\partial \tau} = \frac{\sum_{\mathbf{l}}(e^{-i\mathbf{Q}\cdot\mathbf{l}} - 1)}{n\tau_0} I(Q, \tau) \qquad (10.11)$$

since—

$$\int G_s(\mathbf{r}+\mathbf{l}, \tau)\, e^{i\mathbf{Q}\cdot\mathbf{r}}\, d\mathbf{r} = e^{-i\mathbf{Q}\cdot\mathbf{l}} I_s(Q, \tau)$$

This equation can be integrated by using the boundary conditions of Section 9.5 to give a simple exponential time decay function. Finally Fourier transformation with respect to time yields—

$$S_s(Q, \omega) = \frac{1}{\pi} \frac{f(Q)}{\omega^2 + [f(Q)]^2}$$

(10.12a)

where—

$$f(Q) = -\frac{1}{n\tau_0} \sum_{\mathbf{l}} (e^{-i\mathbf{Q}.\mathbf{l}} - 1)$$

(10.12b)

This equation is similar in form to equation (10.9), but the width of the Lorentzian curve is $c(Q)$ rather than DQ^2.

The problem is now one of expressing $c(Q)$ in a convenient form. Assume first that the vectors $\mathbf{l}$ are oriented at random (i.e. jumping takes place in random directions) and that the distribution of lengths l is a continuous function $a(l)$. In this case, equation (10.12b) is averaged over angles and the sum changed to an integral to give—

$$f(Q) = \frac{1}{\tau_0} \frac{\int \left(1 - \frac{\sin Ql}{Ql}\right) a(l) \, dl}{\int a(l) \, dl}$$

(10.13)

Secondly, if $a(l)$ has the shape of a random distribution of jump lengths [i.e. $a(l) = l \, e^{-l/l_0}$], equation (10.13) reduces to—

$$f(Q) = \frac{1}{\tau_0} \left[1 - \frac{1}{1 + Q^2 l_0^2}\right]$$

(10.14)

It is easy to show from the definition of $a(l)$ that $\overline{l^2} = 6l_0^2$, so that equation (10.12a) reduces to equation (10.9) in the limit of small Q or small l_0.

Equation (10.14) shows that in the limit of large Q, the width of the scattering law (equation 10.12a) is given by $1/\tau_0$. This provides a method of measuring τ_0, if it can be shown that the details of the model are not too important. The principle of this method is illustrated in Fig. 10.3(a) and (b), (the inelastic background of Fig. 10.3(a) corresponds to motion on a time scale $< \tau_0$).

10.4. Mobility and Free Diffusion

The models of the previous two Sections were of the type in which motions at short times were ignored. For example, the expressions for

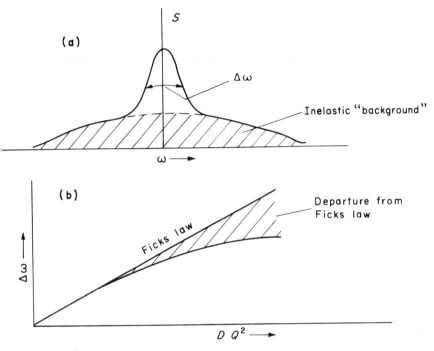

FIG. 10.3. Illustration of measurement of τ_0. $S_s(Q, \omega)$ is measured at constant Q and the width $\Delta\omega$ of the quasi-elastic peak determined (Fig. a). Comparison of $\Delta\omega$ with the predictions of equation (10.13) serves to determine the value of τ_0 (Fig. b). An alternative value of τ_0 obtained by calculating $\Delta\omega$ from equation (10.32) and making another comparison with experiment.

$G_s(\mathbf{r}, \tau)$ did not reduce to the perfect-gas form in the limit of short times or distances. An alternative model can be built up on the basis of the short time behaviour of the atomic motion. As a starting point, the Einstein relation between the diffusion coefficient and the mobility of an atom will be derived.

The mobility (v) is defined by the ratio of velocity ($\mathbf{v}$) of an atom to the force ($\mathbf{F}$) on an atom, i.e.—

$$\mathbf{v}(\mathbf{r}, \tau) = v\mathbf{F}(\mathbf{r}, \tau) \qquad (10.15)$$

If the liquid is in equilibrium, the mean flux of diffusing atoms is zero, i.e.—

$$\mathbf{v}(\mathbf{r}, \tau)G_s(\mathbf{r}, \tau) - D\nabla G_s(\mathbf{r}, \tau) = 0 \qquad (10.16)$$

From a macroscopic point of view, $G_s(\mathbf{r}, \tau)$ may be written in terms of the potential energy, u_s, of the atom at $(\mathbf{r}, \tau)$, so that (compare equation 10.7)—

$$G_s(\mathbf{r}, \tau) = c \exp -\frac{u_s(\mathbf{r}, \tau)}{kT} \qquad (10.17)$$

(where c is a constant).

And since $\mathbf{F} = -\nabla . u_s$ this equation may be rewritten—

$$\nabla . G_s(r, \tau) = -c \exp\left[-\frac{u_s(\mathbf{r}, \tau)}{kT}\right] . \nabla . \frac{u_s(\mathbf{r}, \tau)}{kT}$$

$$= \frac{G_s(\mathbf{r}, \tau) . \mathbf{F}(\mathbf{r}, \tau)}{kT} \qquad (10.18)$$

Comparison of equations (10.16) and (10.18) shows that the mobility is—

$$v = D/kT \qquad (10.19)$$

At this point it is convenient to introduce a model of the atomic motion, and it will be assumed that the atom moves freely for a certain time, then suffers an instantaneous collision after which it moves freely again, but without memory of its previous move. This process, known as free diffusion, is repeated indefinitely and is the "inverse" of jump diffusion (i.e. the time τ_0 is spent in free movement rather than in vibration about a centre of equilibrium). If the elementary time (τ_0) and distance (l) of a diffusive step are used to define the velocity and acceleration appearing in equation (10.15) then it is found that—

$$\tau_0 \simeq M^*v$$

and by using equation (10.19)—

$$= \frac{M^*D}{kT} \qquad (10.20)$$

where M^* is the "effective" mass of the atom. In classical Brownian motion, on which this model is based, the effective mass is just equal to the mass of the particle. However in an actual liquid, the motion of a given atom is impeded by its neighbours and several atoms must be moved in order for diffusion of one to take place. For the purpose of the above discussion, this fact is taken into account by allowing M^* to be much larger than the actual mass. Other definitions of M^* will be used later (Section 12.1).

The behaviour of $G_s(\mathbf{r}, \tau)$ may be deduced from this description of atomic motion. For times small compared to τ_0, the motion will be gas-like and G_s will be given by equation (9.15), whereas for times long compared to τ_0, the random-walk theory will be valid and G_s will be given by equation (10.9). To obtain an expression for $G_s(\mathbf{r}, \tau)$ connecting these two results a more detailed statement of the model is required, and here the description of Brownian motion via the Langevin equation will be used. In this treatment, the atomic motion is supposed to be impeded by frictional forces, and the friction constant is given by—

$$\gamma \equiv \frac{1}{\tau_0} = \frac{kT}{M^*D} \qquad (10.21)$$

This method will be discussed in the next Section.

Before proceeding further a comment is required concerning the relationship between M^* and the length l of a diffusive step. The effective mass for diffusion includes the mass of the atom itself and the mass of neighbouring atoms that have to be moved in order for diffusion to take place. This introduces the idea of a correlation length (denoted by l^*), given by the distance over which some re-arrangement takes place to accommodate the movement of one atom. Thus the correlation length and effective mass are related by the statement that a volume of dimension l^* contains a number of atoms equal to M^*/M. However this length will, in most cases, be significantly larger than the length l for a diffusive step; that is a diffusive step takes place well inside a volume of dimension l^*.

10.5. The Langevin Equation for Brownian Motion

As an illustration of the ideas of the previous Section, the self-correlation function will be calculated from the Langevin equation. This equation is the sum of three terms, the particle acceleration, a damping term involving the friction constant and a driving force, i.e.—

$$\frac{d^2\mathbf{r}}{d\tau^2} + \gamma\frac{d\mathbf{r}}{d\tau} = \frac{1}{M^*}\mathbf{f}(\tau) \qquad (10.22)$$

or—

$$\frac{d\mathbf{v}(\tau)}{d\tau} + \gamma\mathbf{v}(\tau) = \frac{1}{M^*}\mathbf{f}(\tau) \qquad (10.23)$$

where $\mathbf{f}(\tau)$ is a "stochastic" driving force—a force that fluctuates over very short intervals of time, each fluctuation being independent of previous history, and the time average of $\mathbf{f}(\tau)$ is zero. If both sides of equation (10.23) are multiplied by $e^{\gamma t}$ it can be integrated to give—

$$e^{\gamma\tau}\mathbf{v}(\tau) - \mathbf{v}(0) = \frac{1}{M^*}\int_0^\tau e^{\gamma t}\mathbf{f}(t)\,dt \qquad (10.24)$$

A further integration gives—

$$\mathbf{r}(\tau) - \mathbf{r}(0) = \frac{\mathbf{v}(0)}{\gamma}[e^{-\gamma\tau} - 1] - \frac{1}{\gamma M^*}\int_0^\tau dt\,\mathbf{f}(t)[e^{-\gamma(\tau-t)} - 1]$$

$$= \mathbf{A} + \mathbf{B} \qquad (10.25)$$

where $\mathbf{A}$ and $\mathbf{B}$ are a notation for the two terms of this equation. It should be noted that both $\mathbf{A}$ and $\mathbf{B}$ have a normal distribution of zero mean, and are uncorrelated.

The intermediate scattering function for a classical system is obtained by Fourier-transforming equation (8.3). The self term (i.e. $i = j$) is equal to—

$$I_s(Q, \tau) = \langle e^{i\mathbf{Q}\cdot[\mathbf{r}(\tau) - \mathbf{r}(0)]}\rangle \qquad (10.26)$$

The evaluation of equation (10.26) is accomplished with the aid of Blochs theorem, i.e.—

$$\langle e^{i\mathbf{Q}\cdot\mathbf{u}}\rangle = \exp-\tfrac{1}{2}\langle(\mathbf{Q}\cdot\mathbf{u})^2\rangle \qquad (10.27)$$

where $\mathbf{u}$ is a normally distributed variable of zero mean (this is the Gaussian approximation discussed in Chapter 11) Thus combining equations (10.25), (10.26) and (10.27)—

$$I_s(Q, \omega) = \exp-\tfrac{1}{2}[\langle(\mathbf{Q}\cdot\mathbf{A})^2\rangle + \langle(\mathbf{Q}\cdot\mathbf{B})^2\rangle] \qquad (10.28)$$

Now the averages in equation (10.28) are evaluated easily as—

$$\tfrac{1}{2}\langle(\mathbf{Q}\cdot\mathbf{A})^2\rangle = \frac{kT}{2M^*}\frac{(e^{-\gamma\tau} - 1)^2}{\gamma^2}Q^2 \qquad (10.29a)$$

$$\tfrac{1}{2}\langle(\mathbf{Q}\cdot\mathbf{B})^2\rangle = \frac{Q^2}{6}W\left[-\frac{e^{-2\gamma\tau} - 1}{2\gamma^3} + \frac{2(e^{-\gamma\tau} - 1)}{\gamma^3} + \frac{\tau}{\gamma^2}\right] \qquad (10.29b)$$

where—

$$W = \frac{1}{M^{*2}}\int\langle\mathbf{f}(\tau)\cdot\mathbf{f}(\tau + \tau')\rangle\,d\tau'$$

The quantity W may be evaluated directly from equation (10.23) (the cross-term vanishes by the stationarity condition—equation 11.2) as—

$$W = \int\langle\dot{\mathbf{v}}(\tau)\cdot\dot{\mathbf{v}}(\tau + \tau')\rangle\,d\tau' + \gamma^2\int\langle\mathbf{v}(\tau)\mathbf{v}(\tau + \tau')\rangle\,d\tau'$$

$$= 6D\gamma^2 \qquad (10.30)$$

The averages in this equation are given in equations (11.8), (12.12) and (12.16), [note only the long time part of the force correlation is taken as the stochastic part is included in $\mathbf{f}(\tau)$].

It is readily seen that if $\tau \ll 1/\gamma$ equation (10.29b) reduces to zero while equation (10.29a) gives $(kT/2M^*)Q^2\tau^2$ which is the classical gas form. In contrast for $\tau \gg 1/\gamma$ equation (10.29b) gives a term $Q^2 D(\tau - 3/2\gamma)$, whereas (10.29a) gives a constant term $DQ^2/2\gamma$. Thus the limits are as discussed in Section 10.4. However, the actual forms are unsatisfactory, since neither (10.29a) nor (10.29b) is even in τ. Egelstaff and Schofield[37] suggest replacing these expressions by a mathematical model which has the correct limits and is even in τ, i.e.—

$$\tfrac{1}{2}[\langle (\mathbf{Q} \cdot \mathbf{A})^2 \rangle + \langle (\mathbf{Q} \cdot \mathbf{B})^2 \rangle] \equiv DQ^2 \left[\left(\tau^2 + \frac{1}{\gamma^2} \right)^{\frac{1}{2}} - \frac{1}{\gamma} \right] \quad (10.31)$$

Detailed comparison shows that the differences between equations (10.31) and (10.29a) plus equation (10.29b) are minor, but the analytical correctness of equation (10.31) gives I_s a well behaved transform when it is inserted into equation (10.28) i.e.—

$$S_s(Q, \omega) = \frac{\exp DQ^2/\gamma}{\pi} \cdot \frac{DQ^2/\gamma}{[\omega^2 + (DQ^2)^2]^{\frac{1}{2}}} K_1 \left\{ \frac{[\omega^2 + (DQ^2)^2]^{\frac{1}{2}}}{\gamma} \right\} \quad (10.32)$$

where $K_1(x)$ is the Bessel function of the second kind. Since $K_1(x) \to 1/x$ as $x \to 0$ the Lorentzian form (equation 10.9) is recovered if $Q^2 D \to 0$. Also, since $K_1(x) \to \sqrt{(\pi/x)}\, e^{-x}$ as $x \to \infty$, the perfect gas form (equation 9.15) is recovered as $Q^2 D \to \infty$. Comparison of equation (10.32) with experimental measurements of $S_s(Q, \omega)$ provides another method of measuring τ_0(or $1/\gamma$). The method employed is illustrated by Figs. 10.3(a) and (b), and in this case the half-widths are computed numerically from equation (10.32). An example for the case of liquid hydrogen is shown in Fig. 10.4.

10.6. Values of D, τ_0 and l

Two macroscopic methods of measuring D were quoted in Section 10.2. Generally these two methods give data in good agreement and to an accuracy of a few per cent. The time for a diffusive step is measured by comparing neutron-scattering data with, first, equation (10.12a) and, secondly, equation (10.32) using the known value of D. This process gives two estimates of τ_0, which are usually in good agreement, showing that the value obtained is independent of the model. Finally

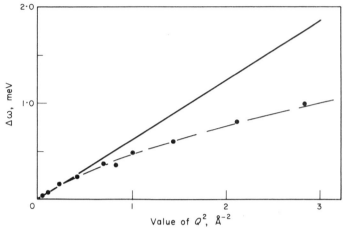

FIG. 10.4. An example of the variation of $\Delta\omega$ (Fig. 10.3a) with Q^2. The full line follows Ficks' Law (with $D = 4\cdot7 \times 10^{-5}$ cm/sec) and the dashed curve is calculated from equation (10.32), with $\tau_0 = 2\cdot6 \times 10^{-12}$ sec. The experimental points for liquid hydrogen at 15°K are shown as full circles. (From Egelstaff, P. A., *et al.*, *Proc. phys. Soc.*, 1967, **90**, 693.)

from the values of D and τ_0 a value of $l \equiv \sqrt{\overline{l^2}}$ is calculated from equation (10.5). A summary of typical values so obtained is given in Table 10.1: the error on the D values is $\sim 2\%$, and the error on the τ_0 values is $\sim 30\%$.

A literal physical interpretation of the jump diffusion and free diffusion models would imply, respectively, that l and M^* are constants. Comparison of the two expressions for τ_0, namely—

$$\left. \begin{aligned} \tau_0 &= \frac{\overline{l^2}}{6D} \text{ (jump)} \\[2mm] \tau_0 &= \frac{M^*D}{kT} \text{ (free)} \end{aligned} \right\}$$
(10.33)

reveals that τ_0 should fall with increasing D for jump diffusion, although it should rise in the case of free diffusion. Table 10.1 shows that there is a tendency for τ_0 to rise for sodium and pentane and fall for hydrogen, yet it is clear that neither the l constant or the M^* constant hypothesis is valid. Consequently the actual diffusive process probably lies between processes described by the two models.

The size of l (Table 10.1) is interesting. On the jump diffusion picture the centre of equilibrium jumps a distance less than one interatomic spacing (especially so in the case of argon) which does not

TABLE 10.1
Experimental values of the diffusion parameters

Liquid	Temperature	$D(10^{-5} \text{ cm}^2/\text{sec})$	$\tau_0(10^{-12} \text{ sec})$	Approx. $l(\text{Å})$
Sodium	108–198°C	4–8	1·0–1·6	2·5
Pentane	−35–+25°C	3–5	1·0–1·2	1·5
Argon	84·5°K	1·6	1	1
Ortho-hydrogen	15–18°K	4·7–7·5	2·6–1·1	2·5

Note: S. J. Cocking (private communication) points out that the value of τ_0 derived from experimental data, via the jump diffusion formula (equation 10.13), depends upon the choice of $a(l)$.

seem to fit the physical idea behind the model. This fact together with the fact that the size of the thermal cloud (~ 0.5 Å) is not small compared to l, confirm that this model is not realistic. On the free diffusion picture, the atom should move freely for 1–2 Å, but this cannot occur in a dense liquid: a difficulty that is supposedly overcome by the idea of an effective mass. However this idea implies that the diffusive steps are not coupled with the other modes of motion (such as those discussed in Chapter 11). Some coupling does exist, as shown in Chapter 11, and so this model is unrealistic also. Thus neither of these models gives a satisfactory description of liquid diffusion, and they must be regarded as mathematical models from which τ_0 may be derived by fitting them to the experimental data. Since they both attempt to cover the same physical situation, namely the behaviour of $G_s(\mathbf{r}, \tau)$ at distances and times comparable to l and τ_0 it is fortunate (and important) that the same values of the parameters are obtained from both models. Moreover, since they represent opposite extremes it may be expected that intermediate models would give similar values too, and for this reason the results in Table 10.1 are believed to be independent of the model.

The above remarks apply to data, such as that illustrated in Fig. 10.3(b), which shows a small departure from Fick's law. If there is a large departure from Fick's law, the details of the model become important, but in this case the quasi-elastic peak (shown in Fig. 10.3a) cannot be unambiguously separated from the inelastic background (shaded in Fig. 10.3a), and the analysis is rather uncertain. A further discussion of this region involves the behaviour of the velocity correlation functions considered in the next Chapter.

The discussion of Chapter 9 suggested that the macroscopic continuum limit was reached for movements over a distance large compared to the interatomic spacing. In this Chapter it was shown that for distances large compared to l, the macroscopic limit is reached. The fact that l (Table 10.1) is of the order of magnitude of the interatomic spacing shows that these two statements are compatible.

Symbols for Chapter 10

$a(l)$	Distribution of jump lengths	$u(\mathbf{r}, \tau)$	Potential energy of atom at $\mathbf{r}$ at time τ
$\mathbf{A}, \mathbf{B}$	Quantities defined by equation (10.25)	W	Time integral of stochastic force
$\mathbf{f}(t)$	Stochastic driving force	v	Mobility of atom
$F(\mathbf{r}, \tau)$	Force on an atom at $\mathbf{r}$ at time τ		

CHAPTER 11

The Velocity Correlation Function

The models given in Chapter 10 are not valid for the short period modes of single particle motion. Short period modes will be discussed in this Chapter, through the velocity correlation function. This function relates the velocity of an atom at a time $\tau = 0$, with its velocity at a later time. It has a simple relation to the diffusion constant (Section 11.1), but it is of physical importance to the liquid state only if the self-correlation function $G_s(\mathbf{r}, \tau)$ is approximately a Gaussian in $\mathbf{r}$. The reason for this is that if G_s is Gaussian, the higher-order velocity correlations (i.e. the correlation of four or more time points) may be shown to be simply related to the "two-time" velocity correlation defined above. Conversely if $G_s(\mathbf{r}, \tau)$ is markedly non-Gaussian in $\mathbf{r}$, the higher-order velocity correlations will have a marked influence on the single-particle modes in addition to the effect included by the simple two time correlation. It is believed that the Gaussian approximation is reasonable,[36] since it is valid in the limits of simple diffusion (equation 10.7), Brownian motion (equation 10.28), the harmonic solid and the perfect gas (equation 9.21). However it is not valid for the jump diffusion model except in the limit of long times. In any real liquid, $G_s(\mathbf{r}, \tau)$ cannot be truly Gaussian in $\mathbf{r}$, and the size of the non-Gaussian terms have been estimated both from experimental measurements of $S_s(Q, \omega)$ and from molecular dynamics calculations to be less than 10–20% of the Gaussian term.

A mathematical statement of the Gaussian approximation has been given in equation (10.27), and this result will be used frequently. The fact that equation (10.27) is equivalent to this approximation may be seen by taking the spacial Fourier transform of a Gaussian function, in which case the right-hand side of equation (10.27) is obtained.

This Chapter will be devoted to a discussion of the two-time velocity correlation and its Fourier time transformation. The latter function (or spectral density) is the analogue of the frequency spectrum of a harmonic solid, and so might be expected to show up the high frequency modes of motion. After discussing the properties of the spectral density and the method of measuring it, two examples of a spectral-density calculation will be given. Finally, these examples will be compared to the experimental data on sodium and argon.

11.1. Relation of Velocity Correlation Function and Diffusion Constant

Imagine an atom moving with a component of velocity $v_x(t)$ along the x direction at the time t. After a time interval $(t_1 - t)$ its velocity will have changed to $v_x(t_1)$ owing to interactions with the body of the liquid. A correlation function, may then be defined as—

$$z(\tau) = \langle v_x(t) . v_x(t_1) \rangle = \langle v_x(0) . v_x(\tau) \rangle \tag{11.1}$$

where $\tau = t_1 - t$. The average in equation (11.1) is over the thermal distribution of $v_x(t)$, and one effect of taking this average is to eliminate the dependence of $z(\tau)$ upon the actual value of t, since at any time each possible initial velocity can be taken. This condition is indicated by the last term in equation (11.1), and is known as the stationarity condition, because—

$$\frac{\mathrm{d}}{\mathrm{d}t} \langle v_x(t) . v_x(t_1) \rangle = 0 \tag{11.2a}$$

or—

$$\langle \dot{v}_x(t) . v_x(t_1) \rangle = - \langle v_x(t) . \dot{v}_x(t_1) \rangle \tag{11.2b}$$

The latter result is widely used in calculations with correlation functions. Also, since the system is in thermal equilibrium the velocity correlation $z(\tau)$ for a classical system must be symmetrical in τ, or—

$$\langle v_x(0) . v_x(t_1 - t) \rangle = \langle v_x(t - t_1) . v_x(0) \rangle \tag{11.3}$$

Now consider the distances L_x which the atom moves along the x axis. The change in position over a long time interval T is—

$$\Delta L_x = L_x(T) - L_x(0) = \int_0^T J(t) \, \mathrm{d}t \tag{11.4}$$

where—

$$J(t) = \frac{\mathrm{d}L_x}{\mathrm{d}t}$$

The mean-square displacement is given by—

$$\overline{(\Delta L_x)^2} = \int_0^T \int_0^T \langle J(t) . J(t_1) \rangle \, \mathrm{d}t \, \mathrm{d}t_1$$

$$= \int_0^T \int_0^T \langle v_x(0) . v_x(\tau) \rangle \, \mathrm{d}t \, \mathrm{d}t_1 \tag{11.5}$$

If the variable of integration t_1 is changed to $\tau = t_1 - t$ and the order of integration is changed by the rule—

$$\int_0^T dt \int_{-t}^{T-t} d\tau \equiv \int_{-T}^0 d\tau \int_{-\tau}^T dt + \int_0^T d\tau \int_0^{T-\tau} dt \equiv \int_0^T d\tau \, 2 \int_\tau^T dt \quad (11.6a)$$

the following result is obtained—

$$\overline{(\Delta L_x)^2} = 2T \int_0^T \langle v_x(0) . v_x(\tau) \rangle \, d\tau \left(1 - \frac{\tau}{T} \right) \quad (11.6b)$$

Finally for spherical symmetry, $\overline{l^2} = 3\overline{(\Delta L_x)^2}$, and taking the macroscopic limit $T \to \infty$, equation (11.6b) becomes—

$$\overline{l^2} = 6T \int_0^\infty \langle v_x(0) . v_x(\tau) \rangle \, d\tau \quad (11.7)$$

A comparison of this equation with equation (10.5) shows that—

$$D = \int_0^\infty \langle v_x(0) . v_x(\tau) \rangle \, d\tau = \int_0^\infty z(\tau) \, d\tau \quad (11.8)$$

This classical result (due to Einstein) is a useful definition of the diffusion coefficient. The validity of the $T \to \infty$ limit has been discussed at length in the literature (e.g. Rice and Gray[5]); it will be assumed without comment here. Equation (11.8) gives one of the averages required in equation (10.30).

In order to understand the relaxation processes that influence the diffusion constant, it is convenient to Fourier-analyse the velocity correlation function. The reason for this is that an oscillator (of frequency ω_0) will give a velocity correlation of the form—

$$z(\tau)_{\text{osc}} = \frac{kT}{M} \cos \omega_0 \tau \quad (11.9)$$

and hence contributes zero to the diffusion coefficient. Thus the spectral density of the velocity correlation is often studied ($\tilde{z}$ denotes a Fourier transform)—

$$\tilde{z}(\omega) = \frac{1}{2\pi} \int_{-\infty}^\infty \langle v_x(0) . v_x(\tau) \rangle \, e^{-i\omega\tau} \, d\tau \quad (11.10)$$

The integral over ω is simply—

$$\int_{-\infty}^\infty \tilde{z}(\omega) \, d\omega = \int_{-\infty}^{+\infty} \langle v_x(0) . v_x(\tau) \rangle \delta(\tau) \, d\tau = \overline{[v_x(0)]^2} = \frac{kT}{M} \quad (11.11)$$

and the limit $\tilde{z}(0) = D/\pi$. In the literature, a function $f(\omega)$ is sometimes employed and is given by $f(\omega) = (2M/kT)\tilde{z}(\omega)$. This function has the property $\int_0^\infty f(\omega)\,d\omega = 1$.

Some simple examples of $\tilde{z}(\omega)$ are shown in Fig. 11.1(a). For a Debye solid $\tilde{z}(\omega)$ is equal to the frequency spectrum and gives a zero intercept at $\omega = 0$, since the diffusion constant is zero. The perfect gas has a constant velocity correlation given by kT/M and thus $\tilde{z}(\omega)$ is equal to $(kT/M)\delta(\omega)$. The Brownian motion example will be considered in Section 11.4. Velocity correlations for the oscillator and for Brownian motion are shown in Fig. 11.1(b).

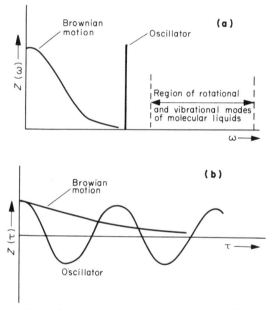

Fig. 11.1. Illustration of velocity correlation function and its spectral density. (a) Spectral density—equation (11.10)—for two models. (b) Velocity correlation—equation (11.1)—for two models.

11.2. The Measurement of $\tilde{z}(\omega)$

The spectral density of the velocity correlation function is obtained as a limit to the self-correlation function of van Hove. As a starting point, the expression for $I_s(Q, \tau)$ in equation (10.26) is rewritten by using Bloch's theorem as—

$$I_s(Q, \tau) = \langle e^{i\mathbf{Q}\cdot[\mathbf{r}(\tau)-\mathbf{r}(0)]}\rangle = e^{-\frac{1}{2}Q^2 w(\tau)}+0(Q^4) \qquad (11.12a)$$

where—

$$w(\tau) = \langle [\mathbf{r}(\tau) - \mathbf{r}(0)]_Q^2 \rangle \qquad (11.12\text{b})$$

and the subscript Q denotes the component along the $\mathbf{Q}$ direction. The terms $0(Q^4)$ arise by separating the distribution of $[\mathbf{r}(\tau) - \mathbf{r}(0)]$ into a normal distribution and a remainder; Bloch's theorem is applied to the normal distribution while the remainder contributes the $0(Q^4)$ terms.

Equation (11.12a) is differentiated twice with respect to time—

$$\frac{1}{I_s(Q,\tau)} \cdot \frac{\partial^2 I_s(Q,\tau)}{\partial \tau^2} = -\frac{Q^2}{2} \frac{\partial^2 w(\tau)}{\partial \tau^2} + \frac{Q^4}{4} \left(\frac{\partial w(\tau)}{\partial \tau} \right)^2 + 0(Q^4) \quad (11.13)$$

which gives the limit—

$$\frac{\partial^2 w(\tau)}{\partial \tau^2} = -\left[\frac{2}{Q^2} \frac{\partial^2 I_s(Q,\tau)}{\partial \tau^2} \right]_{Q\to 0} \qquad (11.14)$$

But using the relation between $I_s(Q,\tau)$ and $S_s(Q,\omega)$—

$$\frac{\partial^2 I_s(Q,\tau)}{\partial \tau^2} = -\int_{-\infty}^{\infty} \omega^2 S_s(Q,\omega)\, e^{i\omega\tau}\, d\omega \qquad (11.15)$$

and combining equation (11.14) with equation (11.15)—

$$\omega^2 \left[\frac{S_s(Q,\omega)}{Q^2} \right]_{Q\to 0} = \frac{1}{4\pi} \int_{-\infty}^{+\infty} \frac{\partial^2 w(\tau)}{\partial \tau^2}\, e^{-i\omega\tau}\, d\tau \qquad (11.16)$$

Now, the expression for $w(\tau)$ can be rewritten—

$$w(\tau) = \langle [\mathbf{r}(t+\tau) - \mathbf{r}(t)]_Q^2 \rangle$$

$$= \text{constant} - 2\langle \mathbf{r}(t+\tau).\mathbf{r}(t) \rangle_Q \qquad (11.17)$$

and hence taking the x direction along Q—

$$\frac{\partial^2 w(\tau)}{\partial \tau^2} = 2\langle v_x(0).v_x(\tau) \rangle \qquad (11.18)$$

The step between equations (11.17) and (11.18) is accomplished by using the stationarity condition (equation 11.2) and differentiating first with respect to t and secondly with respect to τ in order to obtain the result—

$$\langle \dot{r}_x(0).\dot{r}_x(\tau) \rangle = -\langle r_x(0)\ddot{r}_x(\tau) \rangle \qquad (11.19)$$

where dots denote time differentiation. Now the combination of equations (11.18) and (11.16) gives the final result[33]—

$$\omega^2 \left[\frac{S_s(Q, \omega)}{Q^2} \right]_{Q \to 0} = \frac{1}{2\pi} \int_{-\infty}^{\infty} \langle v_x(0) \cdot v_x(\tau) \rangle \, e^{-i\omega\tau} \, d\tau$$

$$= \tilde{z}(\omega) \tag{11.20}$$

This relationship may be used to obtain $\tilde{z}(\omega)$ from measurements of $S_s(Q, \omega)$; although it is valid for a non-Gaussian form of $G_s(\mathbf{r}, \tau)$ it is of practical use only if the $0(Q^4)$ terms in equation (11.12a) do not mar the $Q \to 0$ extrapolation of data taken at a finite Q.

Vineyard[36] has pointed out that the $0(Q^4)$ terms in equation (11.12a) are usually small in comparison to the leading term. This is plausible, since both the small Q (equation 10.9) and the large Q (equation 9.19) expressions meet this requirement. In this case, the scattering law may be calculated from—

$$S_s(Q, \omega) \simeq \frac{1}{2\pi} \int_{-\infty}^{\infty} \exp - \left(\frac{Q^2 w(\tau)}{2} + i\omega\tau \right) d\tau \tag{11.21a}$$

and—

$$w(\tau) = 2\int_0^\tau dt'(t - t') \langle v_x(0) \cdot v_x(t') \rangle \tag{11.21b}$$

The latter equation is analogous to equation (11.6b). Equation (11.21a) is known as the "Gaussian" approximation for S_s, since it leads to a Gaussian function in r for $G_s(r, \tau)$.

11.3. The Moments of $\tilde{z}(\omega)$

It is useful to know the moments of $\tilde{z}(\omega)$ in order to test the models of atomic motion that are often employed to calculate this function. The zeroth moment has been derived at equation (11.11), and since $\tilde{z}(\omega)$ is an even function of ω (because a classical velocity correlation is a real function of τ) the next moment of interest is the second. It is convenient to derive the moments of $\tilde{z}(\omega)$ from the moments of $S_s(Q, \omega)$ and equation (11.20). de Gennes[38] has calculated the classical moments by the following method.

As a starting point, equation (9.9) is employed to derive the following results—

$$2\int_0^\infty \omega^2 S(Q, \omega) \, d\omega = - \left[\frac{\partial^2 I(Q, \tau)}{\partial \tau^2} \right]_{\tau \to 0} \tag{11.22a}$$

$$2\int_0^\infty \omega^4 S(Q, \omega) \, d\omega = \left[\frac{\partial^4 I(Q, \tau)}{\partial \tau^4} \right]_{\tau \to 0} \tag{11.22b}$$

The intermediate scattering function can be written down from equation (9.2) by using expressions (8.5) and (8.6) for $G(r, \tau)$. This yields—

$$I(Q, \tau) = \frac{1}{N} \langle \rho(-Q, 0)\rho(Q, \tau) \rangle \tag{11.23a}$$

where—

$$\rho(Q, \tau) = \int e^{iQ \cdot r} \rho(r, \tau) \, dr = \sum_i e^{iQ \cdot r_i(\tau)} \tag{11.23b}$$

and $\rho(r, \tau)$ is given by equation (8.5).

(Note, this form for $I(Q, \tau)$ is equivalent to equation 10.26).

Equation (11.22a) now becomes—

$$\int_0^\infty \omega^2 S(Q, \omega) \, d\omega = -\frac{1}{N} \left\langle \rho(-Q, 0) \frac{\partial^2}{\partial \tau^2} \rho(Q, \tau) \right\rangle_{\tau \to 0}$$

and using the stationarity condition this becomes—

$$= \frac{1}{N} \left\langle \left| \frac{\partial}{\partial \tau} \rho(Q, \tau) \right|^2 \right\rangle_{\tau \to 0}$$

$$= \frac{Q^2}{N} \sum_{ij} \left\langle \frac{\partial x_i}{\partial \tau} \frac{\partial x_j}{\partial \tau} e^{i(x_i - x_j)Q} \right\rangle \tag{11.24a}$$

where the x axis is along Q and the two positions (i, j) are taken at the same instant of time. In order to evaluate the self term, it is necessary to set $i = j$ in equation (11.24a), so obtaining—

$$2\int_0^\infty \omega^2 S_s(Q, \omega) \, d\omega = Q^2 \langle v_x^2(0) \rangle = \frac{Q^2 kT}{M} \tag{11.24b}$$

Also, the distinct term $(i \neq j)$ vanishes, since the velocities of two atoms taken at the same instant of time are uncorrelated. Thus equation (11.24b) applies to the whole S function and was so employed in equation (9.11). A combination of equation (11.24) with equation (11.20) allows equation (11.11) to be recovered.

Equation (11.22b) when combined with equation (11.23) becomes—

$$2\int_0^\infty \omega^4 S(Q, \omega) \, d\omega = \frac{1}{N} \left\langle \rho(-Q, 0) \frac{\partial^4}{\partial \tau^4} \rho(Q, \tau) \right\rangle_{\tau \to 0}$$

and using the stationarity condition—

$$= \frac{1}{N} \left\langle \left| \frac{\partial^2}{\partial \tau^2} \rho(Q, \tau) \right|^2 \right\rangle_{\tau \to 0}$$

$$= \frac{1}{N} \left[Q^4 \sum_{ij} \langle v_i^2 v_j^2 \, e^{iQ(x_j - x_i)} \rangle \right.$$

$$+ \frac{iQ^3}{M^3} \sum_{ij} \left\langle \left(v_j^2 \frac{\partial U}{\partial x_i} - v_i^2 \frac{\partial U}{\partial x_j} \right) e^{iQ(x_j - x_i)} \right\rangle$$

$$\left. + \frac{Q^2}{M^2} \sum_{ij} \left\langle \frac{\partial U}{\partial x_i} \frac{\partial U}{\partial x_j} e^{iQ(x_j - x_i)} \right\rangle \right] \qquad (11.25)$$

where v_i is the component of velocity along the x axis for the ith atom, and $U\{N\}$ is the potential energy of the N atoms. The acceleration has been eliminated by the relation $M(\partial v_i/\partial \tau) = \partial U/\partial x_i$. The probability of finding N atoms in a configuration of potential energy $U\{N\}$ is $\exp - (U/kT)$, and this gives immediately the Yvon theorem—

$$\left\langle F\{N\} \frac{\partial U\{N\}}{\partial x_i} \right\rangle = kT \left\langle \frac{\partial F\{N\}}{\partial x_i} \right\rangle \qquad (11.26)$$

where $F\{N\}$ is a regular function of the atomic positions.

Selection of the self term $(i = j)$ in equation (11.25) and use of equation (11.26) gives—

$$2 \int_0^\infty \omega^4 S_s(Q, \omega) \, d\omega = 3Q^4 \left(\frac{kT}{M} \right)^2 + \frac{Q^2 kT}{M^2} \left\langle \frac{\partial^2 U}{\partial x_i^2} \right\rangle$$

$$= 3Q^4 \left(\frac{kT}{M} \right)^2 + \frac{Q^2 kT \rho}{M^2} \int_V g(r) \frac{\partial^2 u(r)}{\partial x^2} \, dr \qquad (11.27)$$

where the value of $\langle v_x^4(0) \rangle = 3(kT/M)^2$ for a Maxwellian velocity distribution has been used and the final step results from the pair potential approximation (equation 2.17).

Finally the combination of equations (11.20) and (11.27) gives—

$$\left. \begin{aligned} 2 \int_0^\infty \omega^2 \tilde{z}(\omega) \, d\omega &= \frac{kT}{M^2} \rho \int_V g(r) \frac{\partial^2 u(r)}{\partial x^2} \, dr \\ &= \frac{kT}{3M^2} \rho \int_V g(r) \left[\frac{2}{r} \frac{\partial u(r)}{\partial r} + \frac{\partial^2 u(r)}{\partial r^2} \right] dr \end{aligned} \right\} \qquad (11.28)$$

Equation (11.28) may be used to establish a connection between model parameters and $u(r)$. It is useful also as a means of calculating the frequency range of $\tilde{z}(\omega)$.

11.4. Velocity Correlation Functions for Brownian Motion

As an example of a velocity correlation function consider the model discussed in Section (10.5). Here an atom is imagined to be moving according to the Langevin equation of motion (10.22), and its velocity is given by equation (10.24). To obtain the velocity correlation it is necessary to average over all atoms having the same initial velocity, then multiply through by $v(0)$ and average over the Maxwellian velocity distribution. The term on the right-hand side in $\mathbf{f}(s)$ will disappear provided times of the order of $1/\gamma$ are being considered (i.e. times long compared to the correlation time of the stochastic force) so that $\langle \mathbf{f}(0)\mathbf{f}(t)\rangle = 0$. After completing these steps, equation (10.24) becomes—

$$e^{\gamma \tau}\langle v_x(0) . v_x(\tau)\rangle - \langle v_x^2(0)\rangle = 0$$

or—

$$z(\tau) = \frac{kT}{M}e^{-\gamma \tau} \tag{11.29a}$$

thus—

$$\tilde{z}(\omega) = \frac{1}{\pi}\frac{kT}{M}\frac{\gamma}{\omega^2 + \gamma^2} \tag{11.29b}$$

This form has been drawn in Fig. 11.2.

Now since $\tilde{z}(0) = D/\pi$ it is easy to see that equation (11.29b) gives $\gamma = kT/MD$ in agreement with equation (10.20) if $M = M^*$. However, equation (11.29b) has an infinite second moment in disagreement with equation (11.28), because only the long time effects (i.e. times $\sim \tau_0$) have been included so far. The physical effect which has been omitted is the coupling between the stochastic force $\mathbf{f}(\tau)$ and the friction constant γ (that there is a connection is easily seen by considering the detailed balance condition—or fluctuation dissipation theorem—discussed in Chapter 8). Kubo[39] shows that one way of overcoming this deficiency is to enlarge the definition of the friction constant by making it frequency dependent, and in this case the fluctuation dissipation theorem is satisfied if—

$$\gamma(\omega) = \frac{1}{6MkT}\int_0^\infty \langle \mathbf{f}(0)\mathbf{f}(t)\rangle\, e^{i\omega t}\, dt \tag{11.30}$$

Equation (10.30) is the $\omega = 0$ limit of this equation. If the assumption is made that the stochastic force correlation decays exponentially with constant γ', then—

$$\gamma(\omega) = \frac{1}{MkT}\frac{\langle f^2(0)\rangle}{\gamma' + i\omega} = \gamma_0\frac{\gamma'}{\gamma' + i\omega} \tag{11.31}$$

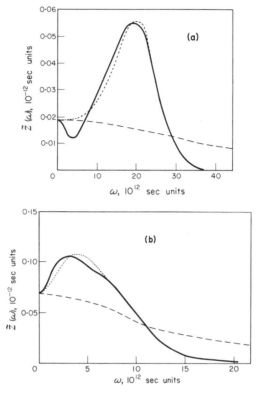

FIG. 11.2. Comparison of the calculated and measured functions $\tilde{z}(\omega)$ for (a) sodium and (b) argon. The full line represents the experimental data, the dashed line equation (11.29b) and the dotted lines equation (11.39). These curves have been normalized to unit area, and in calculating ω_0 from equation (11.40) the integral J_1 was put equal to unity for argon and sodium (for a metal ε and σ are taken from a comparison of the first minimum in $u(r)$—Fig. 4.2—with the L–J potential). The constant, a, equation (11.41) was set equal to 12 for argon and unity for sodium; this difference indicates that the oscillator frequency is better defined for sodium than argon.

where $\gamma_0 = kT/MD$ is the earlier value of γ. Insertion of this value of γ in equation (11.29b) gives a spectral density of the form—

$$\tilde{z}(\omega) = \frac{D}{\pi} \frac{\gamma_0^2 \gamma'^2}{(\omega^2 + \gamma'^2)\omega^2 + \gamma_0^2 \gamma'^2} \qquad (11.29c)$$

This expression has a finite second moment given by (approximately) $(kT/M)\gamma_0\gamma'$. In principle all the higher moments of $\tilde{z}(\omega)$ should be finite which would be so if the above argument were repeated on equation (11.31) and so on. Equation (10.32) gives an exponentially decaying $\tilde{z}(\omega)$ which has this property.

11.5. Velocity Correlation for Brownian Motion of Einstein Oscillators

At the present time it is not possible to calculate the velocity correlation function by a rigorous method, and therefore a model of the liquid is usually employed. In the previous Section the model of Brownian motion was used as an illustration. The comparison with experimental data (Fig. 11.2) shows that it is not a realistic model of a liquid. A more realistic model is the dynamical version of the cell model of Section 5.8. In the cell model each atom is imagined to be confined to a cell and to vibrate as an Einstein oscillator, but a model of this type does not allow for diffusion. To overcome this defect Sears[40] considered a cell model in which the atom is oscillating at a frequency ω_0 about a centre that itself is undergoing Brownian motion. The position of the centre of oscillation at time t is denoted by $\mathbf{r}(t)$ and is defined by—

$$\mathbf{r}(t) = \frac{1}{t_0} \int_{t-t_0/2}^{t+t_0/2} \mathbf{r}_0(t')\, dt' \qquad (11.32)$$

where t_0 is the period of oscillation (i.e. $2\pi/\omega_0$) and $\mathbf{r}_0$ denotes the position of the atom whose velocity correlation is to be calculated. The motion of this atom (called the central atom) is affected by the positions and motions of surrounding atoms (which form the cell), and the effect produced by these atoms is to be calculated. The following notation will be used for the displacements i.e.—

$\mathbf{a}_j(t) = \mathbf{r}_j(t) - \mathbf{r}(t)$ = position of atom j relative to the centre of oscillation.

$\mathbf{b}(t) = \mathbf{r}_0(t) - \mathbf{r}(t)$ = thermal vibration displacement of the central atom.

The total force on the central atom may be expanded in powers of $\mathbf{b}(t)$, i.e.—

$$\mathbf{F}(t) = M\mathbf{A}(t) - M\omega_0^2 \mathbf{b}(t) - \mathbf{K}(t) \cdot \mathbf{b}(t) + 0(b^2) \qquad (11.33)$$

where $M\mathbf{A}(t)$ is a fluctuating driving force (with average value zero) arising from the thermal motion of atoms in the neighbourhood of the central atom, $-M\omega_0^2 \mathbf{b}(t)$ is the average restoring force on the central atom, $-\mathbf{K}(t) \cdot \mathbf{b}(t)$ is the modulation of the restoring force by the thermal motion of the atoms near the central atom and $0(b^2)$ represents the anharmonic part of the restoring force. Owing to the symmetry requirements, it follows that—

$$\langle A_\alpha(t) \rangle = 0; \qquad \langle A_\alpha(t) \cdot A_\beta(t) \rangle = \delta_{\alpha\beta} \langle A_\alpha^2 \rangle$$

and—

$$\langle K_{\alpha\beta}(t) \rangle = 0$$

where α, β are Cartesian indices. In terms of the pair potential, $u(r)$, it is straightforward to show that—

$$\omega_0^2 = \frac{1}{M} \sum_i \left\langle \frac{\partial^2 u(a_i)}{\partial a_{i\alpha}^2} \right\rangle$$

$$= \frac{\rho}{M} \int_V g(a) \frac{\partial^2 u(a)}{\partial a_\alpha^2} \, da \qquad (11.34a)$$

and—

$$A_\alpha(t) = \frac{1}{M} \sum_i \frac{\partial u(a_i)}{\partial a_i} \cdot \frac{a_{i\alpha}}{a_i}$$

so that—

$$\langle A_\alpha^2(t) \rangle = \frac{\rho}{3M^2} \int_V \left[\frac{\partial u(a)}{\partial a} \right]^2 g(a) \, da$$

$$+ \frac{\rho^2}{3M^2} \int\int_V \frac{\partial u(a_1)}{\partial a_1} \frac{\partial u(a_2)}{\partial a_2} g_3(a_1 a_2) \frac{\mathbf{a}_1 \cdot \mathbf{a}_2}{a_1 a_2} \, d\mathbf{a}_1 d\mathbf{a}_2$$

$$(11.34b)$$

At this stage it is necessary to introduce several approximations. The first is to regard $A(t)$ and $K(t)$ as stochastic variables, and this eliminates those physical effects which dissipate the thermal vibration energy of the central atom into the surrounding liquid. To compensate for this, a dissipative force proportional to $\dot{r}_0(t)$ is introduced. In addition the frequency modulation term $K(t) \cdot \mathbf{b}(t)$ has a dissipative effect upon the oscillations and so it is arbitrarily included in the former term. Finally the anharmonic terms are neglected since b^2 is assumed to be small, giving the equation—

$$M\ddot{r}_0(t) = \mathbf{F}(t) = M A(t) - M\omega_0^2 [r_0(t) - r(t)] + M\mu_0 \dot{r}_0(t) \qquad (11.35)$$

where μ_0 is an arbitrary friction constant. As in Section (11.4), the random force and the friction constant must be connected by equation (11.30), and this modifies equation (11.35) to—

$$\ddot{r}_0(t) + \int_0^t \mu(t - t') \dot{r}_0(t) \, dt' + \omega_0^2 \{r_0(t) - r(t)\} = A(t) \qquad (11.36)$$

which is the equation for a damped Einstein oscillator. Finally, the motion of the centre of oscillation is assumed to be governed by the Langevin equation (including 11.30) which has the form—

$$\ddot{r}(t) + \int_0^t \gamma(t - t') \dot{r}(t') \, dt' = \frac{1}{M} \mathbf{f}(t) \qquad (11.37)$$

It should be noted that the actual mass is used in these equations rather than M^*.

The coupled equations (11.36) and (11.37) can be solved by a method analogous to that described earlier for the simple Langevin equation. To simplify the result, it will be assumed here that the friction constant affecting the stochastic forces is the same in both equations, or $\mu' = \gamma'$. Also from $\tilde{z}(0) = D/\pi$, it can be shown that—

$$\frac{\gamma'}{\langle f_\alpha^2 \rangle} = \left(\frac{M}{kT}\right)^2 D \quad \text{or} \quad \frac{M}{kT}\langle f_\alpha^2 \rangle = \mu'\gamma_0 \tag{11.38}$$

and in this case the solution[41] is—

$$\tilde{z}(\omega) = \frac{D}{\pi}\left[\frac{\gamma_0^2\mu'^2}{(\omega^2+\mu'^2)\omega^2+\gamma_0^2\mu'^2} + \frac{\gamma_0\mu'}{\omega_0^2}\cdot\frac{\omega^2 a}{\omega^2+\mu'^2}\right]$$

$$\times \left[\left(\frac{\omega^2-\omega_0^2}{\omega_0^2}\right)^2 + \frac{\omega^2 a^2}{\omega^2+\mu'^2}\right]^{-1} \tag{11.39}$$

and μ' may be determined from $\int \tilde{z}(\omega)\,d\omega = kT/M$. Also (see below) the quantity $M/kT\langle A_\alpha^2\rangle$ has been replaced by $a\omega_0^2$, and γ_0 is defined at equation (11.31).

Sears[40] evaluates equations (11.34a and b) for a L–J potential, obtaining—

$$\left.\begin{aligned}\omega_0^2 &= \frac{32\pi\varepsilon\sigma\rho}{M}J_1 \\[2mm] \langle A_\alpha^2\rangle &= \frac{64\pi\varepsilon^2\sigma\rho}{3M^2}J_2\end{aligned}\right\} \tag{11.40}$$

where J_1 and J_2 are integrals that have a numerical value of approximately unity. Near the triple point, $\varepsilon \sim 1\cdot5\,kT$ (Table 1.1), and with this approximation—

$$\langle A_\alpha^2\rangle = a(kT/M)\omega_0^2 \tag{11.41}$$

where a, is of the order unity. In this case, the model is specified by the pair potential well (i.e. the parameters ε and σ) and by the diffusion constant D, and the lifetime of the oscillator is related to the parameter a.

The interpretation of this model has not been worked out in detail so far: for example, there are two views on the value of γ_0 to be used in the first term in the brackets of equation (11.39). On the simple Brownian model, it is proportional to $1/M$; however, from the physical significance of this term (i.e. the diffusion term) it should be proportional to $1/M^*$ (i.e. to $1/\tau_0$). Also the definition of ω_0^2 (equation

11.34a) implies that the second moment of $\tilde{z}(\omega)$ is equal to $(kT/M)\omega_0^2$: it is not yet clear that this result is consistent with the zero moment in every case. Consequently a further weak restriction may be placed on the parameter (which may be covered by relaxing the approximation $\mu' = \gamma'$). Finally there is some uncertainty about the way in which the fluctuation–dissipation theorem has been used in the coupled equations (11.36) and (11.37). In spite of the above difficulties the basic results of the model may be expected to be valid. For the examples given in the next Section the Sears[40] approximation $\gamma_0 = (kT/MD)$ will be used, and the other difficulties will be neglected.

11.6. Comparison of Theoretical and Experimental Values of $\tilde{z}(\omega)$

The shape of $\tilde{z}(\omega)$ for sodium has been obtained in neutron-scattering experiments[42] and that for argon has been calculated by an atomic dynamics procedure.[43] In both cases, the simple Brownian motion model may be shown to be inadequate (Fig. 11.2) and a variety of more complex models have been advanced. All involve some attempt to account for the quasi-oscillatory motion of the atoms, which modifies the large ω part of $\tilde{z}(\omega)$. There is no substantial difference between the various methods, and in this sense the Sears model (Section 11.5) is typical. Comparison between the experimental and model calculations for sodium and argon are shown in Figs. 11.2(a) and (b). The agreement between theory and experiment is good, although in both cases the calculated curve shows a slightly sharper peak than observed in practice. This effect is almost certainly due to the assumption of a single oscillator frequency and an arbitrary frequency distribution could be introduced to improve the fit. However, such a modification is probably unjustifiable in view of the weaknesses of the model, particularly since the width of the peak in $\tilde{z}(\omega)$ has been adjusted through the constant a. The difference between the values of a (12 compared to 1) for argon and sodium probably reflects the difference in shape of the potential wells shown in Figs. 3.1(a) and 4.2, since the sharper well for a metal probably leads to a better definition of the oscillator frequency.

The two models discussed (Brownian motion, equation 11.29c, and the cell oscillator, equation 11.39) can be taken as (opposite) extreme views of the function $\tilde{z}(\omega)$. In the former model, the oscillator modes are completely damped, so that only the random Brownian motion survives. In the latter model, the assumption of a single oscillator over-emphasizes the vibrational characteristics of the system when the Brownian motion effect is small. Nevertheless, these models do demonstrate the type of effects entering $\bar{z}(\omega)$ and the information needed to calculate $\bar{z}(\omega)$, and as the experimental data is improved will provide the key to more sophisticated calculations. One of the features that is ignored by these models is coupling between the low-frequency diffusive modes and the high-frequency modes—the Sears model, for example, is based on the assumption of no coupling. Coupling of this type is an important physical property of the liquid that must be included in a complete treatment of diffusion. The minimum in the data on sodium (Fig. 11.2a) suggests that a slight division occurs between these two regions, but this feature of the curve is just significant only (owing to the size of the experimental error). Further experimental and theoretical work is required to establish the magnitude of the coupling.

Symbols for Chapter 11

$A(t)$	Fluctuating driving force in equation (11.33)	$K(t)$	Modulation of restoring force in equation (11.33)
a	Constant in equation (11.41)	L_x	Distance along x axis
$\left.\begin{array}{l}\mathbf{a}_j\\ \mathbf{b}\end{array}\right\}$	Position parameters in itinerant oscillator models	T	A long time
		ω_0	Oscillator frequency
$F(t)$	Total force on atom in equation (11.33)	μ	Friction constant
		μ'	Time decay constant for $A(t)$
$\mathbf{J}(t)$	Particle current at time t		

CHAPTER 12

Phenomenological Treatments of Diffusion and Viscosity Coefficients

It will have been noticed that the diffusion constant was expressed in terms of velocities, distances and times rather than in terms of the pair potential and the pair correlation function. The method of calculation adopted in the last two Chapters has differed from that used earlier, in that, unlike Chapters 2, 5 and 7, statistical mechanics was not used extensively. This points up a basic difference between methods of calculating the equilibrium properties and the transport properties. As an illustration of these differences, Fig. 12.1 shows the main steps in the mathematical procedure. For equilibrium properties, classical mechanics and the pair potential is the starting point and the calculation proceeds via equilibrium statistical mechanics. Along this route the information regarding time variation of positions is lost. In contrast, for transport coefficients the calculation proceeds via a detailed molecular dynamics method (for the monatomic systems such as argon and sodium, this would be an atomic dynamics calculation). Such a calculation may be accomplished in one of two ways; either the equations of motion for each atom may be written down and solved continuously on a large computer, or a model for the liquid may be assumed and the dynamical consequences of the model worked out. Several examples of the latter method have been given in Chapters 10 and 11.

This difference may be described in another way. It was shown in Chapter 8 that the pair correlation function $g(r)$ was the $\tau = 0$ limit of the time dependent correlation function. Similarly, the quantities that appear in the dynamical models of the transport coefficients and which depend upon $g(r)$ may be shown to be $\tau = 0$ limits of physical properties (e.g. equation (11.28) for the width of $\tilde{z}(\omega)$ may be shown via the equation (12.10) to give the $\tau = 0$ value of the force correlation). But the transport coefficients themselves are integrals over all time as shown by equation (11.8) for example. These integrals cannot be expressed in a simple way in terms of the pair potential and pair

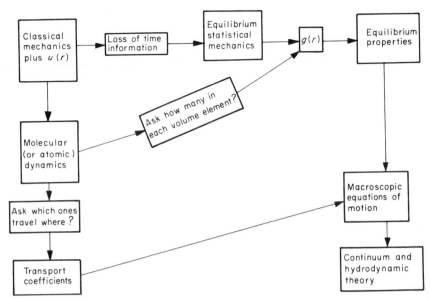

Fig. 12.1. Block diagram showing routes for the calculation of equilibrium and transport properties. The manner in which these results fit into continuum theory is indicated also.

distribution functions, and so cannot be calculated in a similar way to the other properties. Further examples of dynamical models used to calculate transport coefficients will be given in this Chapter.

In the continuum or hydrodynamic region of (ω, Q) space macroscopic equations of motion are used to calculate the behaviour of the liquid. For such a theory the values of the equilibrium properties and of the transport coefficients are employed as given constants. This method is shown on the right-hand side of Fig. 12.1. These equations of motion provide a theoretical basis for discussing the macroscopic behaviour of liquids, but are dependent upon a set of unknown parameters. If the microscopic theory is successful it will be able to predict the values of these parameters and in this sense the continuum theory depends upon the microscopic theory for its success. At the present time, accurate theoretical values of the equilibrium properties and the transport coefficients are not available and therefore experimental values of the parameters are normally employed in the continuum theory. In Chapter 13, the propagation of waves through a continuum will be discussed on this basis. It will be shown there again how the constants which are related to the $\tau = 0$ limit can be given in terms of $g(r)$ and $u(r)$, and the coefficient of viscosity, for example, is given by a time integral and may not be described simply in this way.

Consequently, progress in understanding the transport coefficients depends upon phenomenological analysis. In this Chapter a variety of attempts of this kind will be described, each of which tends to emphasize one particular dynamical feature. These methods are based on models of the dynamical behaviour of the liquid and are illustrations of the left-hand route shown in Fig. 12.1.

12.1. Problems in the Calculation of the Diffusion Coefficient

The calculation of the diffusion coefficient in terms of the pair potential and pair distribution functions has not been successful so far, except in certain limiting cases [e.g. through the use of equation 11.28 and a simple form for $\tilde{z}(\omega)$]. The reason is, essentially, that in equation (11.8) many Fourier components of the velocity give a zero contribution. Another way of making this point is to say that many degrees of freedom are not concerned with diffusion, so that the behaviour of $\tilde{z}(\omega)$ near $\omega = 0$ is of interest rather than the whole range of ω. The idea of an effective mass (M^*) for diffusion is of interest here since the area of $\tilde{z}(\omega)$ is proportional to $1/M$, and a feature in the spectrum $\tilde{z}(\omega)$ may be given an effective mass proportional to the reciprocal of its area. Thus if each Fourier component of $\tilde{z}(\omega)$ is associated with a number of degrees of freedom proportional to its area, the effective mass will be proportional to the reciprocal of the number of degrees of freedom. If only a small region of $\tilde{z}(\omega)$ is associated with the diffusive processes, then the effective mass for diffusion will be much larger than the actual mass. The experimental $\tilde{z}(\omega)$ for sodium in Fig. 11.2(a) has a relatively small peak near the origin which is associated with diffusive processes. For argon (Fig. 11.2b) the situation is more complex and a larger region of $\tilde{z}(\omega)$ seems to be associated with diffusion. However in both cases the high-frequency part of $\tilde{z}(\omega)$ is associated with other processes, and hence there must be substantial cancellation in finding the area of $z(\tau)$ in equation (11.8).

For this reason it is usual to propose a model of the liquid and then calculate the diffusion constant for the model. In many cases, the model will be based on phenomenological parameters. Broadly, this situation holds for the viscosity[44] coefficient too, and since on a simple view (Section 12.2) the two coefficients are related a selection of the models for both coefficients will be discussed in this Chapter. It will be concluded that there is so far, no clear solution to the above problem.

12.2. Qualitative Relation between Diffusion and Viscosity Coefficients

If each atom in the liquid is sitting in a potential well, then for a movement from one site to another it is necessary first for a vacancy to be formed and secondly for an atom to jump over a potential barrier into the vacant site. This is illustrated by Fig. 12.2(a). The energy W is the sum of the energies to form the vacancy and to clear the potential barrier. Thus the chance to jump from one site to another is proportional to $e^{-W/kT}$ and this must be proportional to the diffusion constant or—

$$D = D_0\, e^{-W/kT} \tag{12.1}$$

where D_0 is a constant. A system following a relationship of this type is said to show "Arrhenius" behaviour.

In contrast the shear viscosity will be related to the time an atom remains at a given site, since in a dense liquid it is governed by the rate at which shear can take place between two layers of atoms. This is illustrated in Fig. 12.2(b). Thus the viscosity is proportional to the reciprocal of the chance for a jump from site to site, or to $e^{W/kT}$. However according to this point of view, the viscosity will be proportional to $1/v$ (where v is the mobility), and hence from equations (10.19) and (12.1)—

$$\eta = \eta_0 \frac{T}{T_0} e^{W/kT} \tag{12.2}$$

where η_0 is the viscosity at some high temperature T_0.

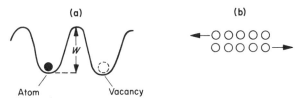

| (a) | (b) |

Atom   Vacancy

FIG. 12.2. Phenomenological picture of diffusion and viscosity coefficients. (a) Diffusion coefficient is proportional to the chance to jump from site to site. (b) The viscosity coefficient is related to the lifetime at each site [i.e., 1/(chance to jump)].

Figure 12.3 shows the measured diffusion and viscosity coefficients of argon and sodium plotted on semi-logarithmic scales against temperature. A smooth line has been drawn through to actual experimental points to illustrate the fact that they satisfy equations (12.1) and (12.2) approximately. The values of W are listed in Table 12.1.

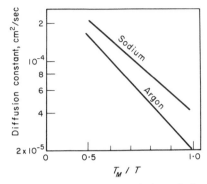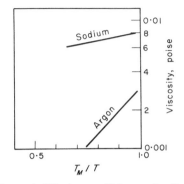

FIG. 12.3. Temperature variation of viscosity and diffusion coefficients of sodium and argon (T_M is the melting temperature). A smooth line has been drawn through the actual data to illustrate the approximate behaviour.

TABLE 12.1
Values of W/kT close to the melting point at 1 Atm

Liquid	A	CH_4	Na	Hg
W/kT from D	3·7	4·6	3·3	2·5
W/kT from η	3·2	3·0	2·2	1·4

There is a rough correspondence between the two values of W/kT, although the diffusion constant values are the greater. In addition, it is found that the temperature dependence is not exactly exponential, which is sometimes expressed by saying that W varies with temperature. Alternative forms for the temperature dependence are often tested and, for example, it is found that for many metals a linear variation with temperature fits as well (or sometimes better) than an exponential.

It is interesting to compare the values of W for the solid and liquid states. For crystals it is found that W is of the order of the latent heat of vaporization (for solid sodium near the melting point $W/kT \sim 12$). In contrast, for liquids W is about 0·2 of the latent heat of vaporization. Sometimes[46] this difference is explained by referring to the equation—

$$p_j \simeq p_E p_V \qquad (12.3)$$

where p_j is the jump probability, p_E is the probability of finding sufficient energy to break the bonds between neighbours and p_V is the probability of finding the necessary "free volume" to make the move. The difference in "free volume" (i.e. the volume not occupied by atoms at a given instant) between the solid and liquid states is then assumed to lead to large differences in p_V. Cohen and Turnbull,[47] for example, show that—

$$p_V \propto \exp - V/V_f \tag{12.4}$$

where V_f is the free volume and on certain assumptions it can be proportional to T. In this case equation (12.3) will give the same temperature variation as equation (12.1).

12.3. Rate Theory Connection between Diffusion and Viscosity Coefficients

Eyring (see ref. 46) developed a theory of rate processes which he employed to connect the diffusion and viscosity coefficients. Figure 12.4 is an enlargement of Fig. 12.2(b) showing the viscous shearing process in more detail. It is assumed that in order for an atom to move from point A to point B it must pass over a potential barrier which is symmetrical in the absence of an external force. However if an external force, **F**, is applied then the barrier will not be quite symmetrical as shown in the figure. The frequency of jumps in the direction of the applied force (k_f) will be greater than the frequency in the reverse direction (k_r). Thus the net velocity (v) of flow with respect to the sites is—

$$v = a(k_f - k_r) \tag{12.5}$$

where a is the spacing between atoms.

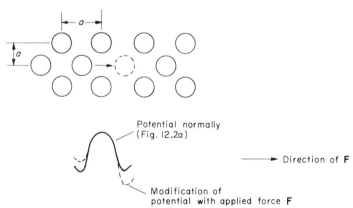

Potential normally (Fig. 12.2a)

Direction of **F**

Modification of potential with applied force **F**

FIG. 12.4. Illustration of viscous shearing process when a force **F** is applied to the liquid. The circles represent atoms of spacing, a, and the dotted circle represents a vacancy. In the lower part of the figure the modification in the potential due to the application of the force is shown. (From Glasstone et al.,[46] Figs. 118 and 119.)

Now the force acting on a particular atom is $F/\rho a$ (where ρ is the number density and F is the force per unit area), and this force acts so as to decrease the activation energy in the direction of $\mathbf{F}$. If the activated state is halfway between the two sites then the work done in moving the molecule $\frac{1}{2}a$ is $F/2\rho$ and the frequencies may be written—

$$k_f = k_0\,e^{F/2\rho kT}\,; \qquad k_r = k_0\,e^{-F/2\rho kT} \tag{12.6}$$

where k_0 is the rate in the absence of a force. From the definition of shear viscosity (namely the ratio of the tangential stress to the shear rate), it is seen that—

$$\eta = \frac{aF}{v} = \frac{F}{2k_0\,\sinh(F/2\rho kT)}$$

$$\simeq \frac{\rho kT}{k_0} \tag{12.7}$$

The first step involves equations (12.5) and (12.6), and the second is valid for small forces F.

Now k_0 is related to the diffusion coefficient as follows. Assume that ρ is a function of position (x); then the difference in density between neighbouring layers is $a(d\rho/dx)$. The number of atoms that pass from unit area of the first layer to the second layer in unit time is $a\rho k_0$, and hence from the definition of the diffusion constant—

$$ak_0\!\left(a\frac{d\rho}{dx}\right) = D\frac{d\rho}{dx} \quad \text{or} \quad D = a^2 k_0 \tag{12.8}$$

The combination of equations (12.8) and (12.9) gives—

$$\eta = \frac{kT}{D}a^2\rho = \frac{kT}{aD} \qquad (\text{if } \rho = a^{-3}) \tag{12.9}$$

This relationship is similar to equation (10.20), and shows that in this model the shear viscosity is equivalent to the friction constant, and the distance a has to be related to the correlation length rather than the size of one atom.

12.4. The Force Correlation Method

Since (equation 10.20) the friction constant is related to the diffusion constant, it is also, by equation (12.9) closely associated with the shear viscosity. For this reason, there have been several attempts to calculate its value for simple liquids, and some of these methods start from the force correlation function. By differentiating the velocity correlation twice with respect to time and using the stationarity condition, it may be readily shown that—

$$F(\tau) = \langle K(0) . K(\tau) \rangle = M^2 \frac{d^2}{d\tau^2} \langle v_x(0) . v_x(\tau) \rangle \qquad (12.10)$$

where $K(\tau)$ is the force in the x direction on the atom at time τ. The force correlation for $\tau = 0$ may be derived from equation (11.28)—

$$\langle K^2(0) \rangle = \frac{\rho k T}{3} \int_V \nabla^2 u(r) g(r) \, d\mathbf{r} \qquad (12.11)$$

and the area of the force correlation function vanishes—

$$\int_0^\infty \langle K(0)K(\tau) \rangle \, d\tau = 0 \qquad (12.12)$$

The latter point follows from the fact that the Fourier transform of equation (12.11) is—

$$\tilde{F}(\omega) = M^2 \omega^2 \tilde{z}(\omega) \qquad (12.13)$$

so that $\tilde{F}(0) = 0$.

Two of the equations used to estimate the friction constant are the Kirkwood[48] equation—

$$\gamma = \frac{1}{MkT} \int_0^\infty \langle K(0)K(\tau) \rangle \, d\tau \qquad (12.14)$$

and the Kirkwood and Rice[49] equation—

$$\gamma^2 = \frac{\rho}{3M} \int_V \nabla^2 u(r) g(r) \, d\mathbf{r} \qquad (12.15)$$

[this equation implies that γ^2 is the mean square value of ω over $\tilde{z}(\omega)$—compare equation (11.28)]. If these equations are compared to equations (12.12) and (12.11), respectively, it is clear that they cannot be applied to the whole of the force correlation function. In practice they have to be applied to the long time part of it and thus some "arbitrary" selection between different regions has to be made. For this reason, the difficulty discussed in Section 12.1 arises in this method too. The necessary selection is illustrated by applying equation (12.10) to the Langevin velocity correlation (equation 11.29), when—

$$\langle K(0)K(\tau) \rangle = MkT\{2\gamma\delta(\tau) + \gamma^2 e^{-\gamma\tau}\} \qquad (12.16)$$

The first term on the right-hand side represents the stochastic driving force in this problem, and the second term represents the force acting over a long time. This term may be used in equations (12.14) and (12.15). Equation (12.15) gives in effect the square of the width of the function $\tilde{z}(\omega)$, and will only give the correct friction constant if $\tilde{z}(\omega)$ is wholly due to diffusion processes. This is illustrated by equation (11.34a), which shows that the width of $\tilde{z}(\omega)$ is governed by the frequency of oscillations (if present) rather than the diffusion process.

Several authors have elaborated the latter point by assuming that the oscillatory behaviour is associated with co-operative modes of motion. As an example the result of Rice[50] will be quoted—

$$\gamma_s = \frac{\rho}{36\pi M^2 c^3} \left[\int_V \nabla^2 u(r)g(r)\,d\mathbf{r} \right]^2 \qquad (12.17a)$$

where c is a propagation velocity. Rice concludes that γ_s is only part of the friction constant and a second part (due to hard-core interactions that are excluded from equation 12.17a) must be added. For a hard-sphere potential, this term is—

$$\gamma_h = \tfrac{8}{3}\rho\sigma^2 g(\sigma)\sqrt{\frac{\pi k T}{M}} \qquad (12.17b)$$

The sum $\gamma = \gamma_s + \gamma_h$ is then used to calculate the diffusion constant and the result (for argon) is given in Fig. 12.5. A reasonably good fit is obtained, but this clearly depends upon the choice of $g(\sigma)$, which is fairly arbitrary, and upon the evaluation of the integral in equation (12.17a), which is difficult to do accurately (Chapter 7).

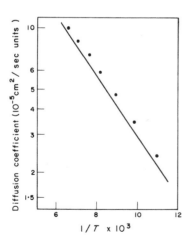

FIG. 12.5. Diffusion constant of argon calculated from equation (12.17) using approximate distribution functions and c = velocity of sound—continuous line; to be compared to experimental values—full circles, (From S. A. Rice, *In* "Liquids—Structure Properties and Solid Interactions" (Ed. T. J. Hughel), Fig. 12, p. 134.)

Recent work (Rice and Gray,[5] p. 462) indicates that c should be the velocity for high-frequency modes (Chapter 14) and that this value of c with more accurate distribution functions gives a line similar to that shown in the Figure.

12.5. Discussion of Length and Time of a Diffusive Step

It has been pointed out that the diffusion constant near the triple point is about 10^{-5} cm^2/sec for many liquids. The physical reasons for this are not fully understood, but some light can be thrown on this problem by noting that if the diffusion constant is lowered below this limit, the liquid freezes and becomes a solid. Because of this, the problem can be inverted to "what is the highest value of D for a solid?". In order to answer this question, the values of l and τ_0 in equation (10.5) are required. The choice of l for a crystalline solid has to be close to the interatomic spacing or about 3 Å. The time for a diffusive step is more difficult to assess, but since it corresponds to the "memory time", it may be associated with the life-time for lattice vibrations. There are two aspects to this problem illustrated by Figs. 12.6(a) and (b). One aspect concerns the time for an atom to build up a thermal cloud. Or if an atom is placed on a particular site, how long is it before lattice vibrations have smeared out its position— that is before its energy has been shared with other atoms? This question is illustrated in Fig. 12.6(a) and the time involved is $\sim 10^{-12}$ sec. Another time is the life-time of a phonon at a temperature near to the melting point of a crystal lattice (see Fig. 12.6b). This may be calculated roughly from anharmonic theory and may be measured by inelastic neutron scattering methods. Again this time is $\sim 10^{-12}$ sec for a phonon of reasonably high frequency; such phonons should be considered, since they carry most of the thermal energy. Consequently, the shortest τ_0 (and hence the largest D) that can be considered for the lattice is $\sim 10^{-12}$ sec, and the combination of these values of l and τ_0 give a maximum D for a lattice of $\sim 10^{-5}$ cm^2/sec. In practice, crystal lattices melt before D has reached this value, because it

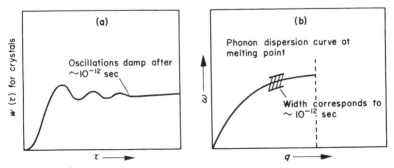

Fig. 12.6. Illustration of memory time in a crystal lattice. (a) Behaviour of $w(\tau)$. (b) Phonon life time.

implies an almost continual state of diffusion, i.e. the atom gathers sufficient thermal energy to make a jump, then after the jump it distributes this excess energy among its neighbours, but almost immediately it gathers sufficient energy for another jump. Thus a crystal will melt for $D \sim 10^{-5}$ cm²/sec, and thus the simplest view to adopt is that a liquid has a $D \gtrsim 10^{-5}$ cm²/sec.

This difference between the solid and the liquid can be observed directly in the function $S_s(Q, \omega)$. The period of vibration of an atom in a crystal lattice is $\sim 10^{-13}$ sec, or at most $\frac{1}{10}$th (usually $< \frac{1}{1000}$th) of the time for a diffusive step. Thus the atom vibrates about a (lattice) site for a relatively long time before jumping, and this defines the lattice. In contrast, the diffusion rate in a liquid is sufficiently fast that it is difficult to distinguish between oscillations and jumps. Thus the motion is more continuous than in the solid, and this means that the elastic and inelastic regions of $S_s(Q, \omega)$ become merged together. These differences are illustrated in Fig. 12.7(a) and (b) for the liquid and solid, respectively.

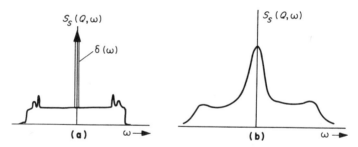

FIG. 12.7. Illustration of shape of $S_s(Q, \omega)$—for $Q \sim \frac{1}{2}Q_0$—in the cases of: (a), a crystal; (b), a liquid.

The value of τ_0 may be shown to be $\sim 10^{-12}$ sec from the values of the diffusion and viscosity coefficients, that is without assuming a value for l or a crystalline model. To do this, the correlation length (l^*) is calculated from equation (12.9), assuming that $l^* \equiv a$, and the effective mass is calculated by assuming it is the mass of a sphere of diameter l^*, i.e.—

$$M^* = M\frac{\pi l^{*3}\rho}{6} \qquad (12.18)$$

(note: for equation 12.9, ρ is the number density of spheres of diameter $l^* = a$; here ρ is the actual number density). Then from equation

(10.20) the effective mass is related to τ_0 giving—

$$\tau_0 = \frac{MD}{kT} \cdot \frac{\pi l*^3 \rho}{6} \tag{12.19}$$

As stated above for liquids near the triple point, equation (12.19) gives $\tau_0 \sim 10^{-12}$ sec.

However the above estimates of τ_0 do not suggest that its temperature dependence will be as small as indicated in Table 10.1. If $\tau_0 D$ increases with temperature—as found experimentally in some cases—then l must increase with temperature too. From Table 10.1, this increase is faster than the increase in (density)$^{-\frac{1}{3}}$ showing that interatomic spacing is not the controlling factor. The significance of this result is not understood.

12.6. Connection between Viscosity and Thermal Conductivity

The diffusion, viscosity and thermal conductivity (λ) coefficients for a dilute hard sphere gas are simply related[10] as—

$$D = \frac{\eta}{M\rho} \tag{12.20a}$$

$$\lambda = \frac{C_V \eta}{M} \tag{12.20b}$$

equation (12.20a) is invalid for a liquid, since it is based upon the assumption that the momentum transport takes place through particle transport. This assumption implies that the system does not permit co-operative modes of motion. Equation (12.20b), however, has been found in practice to apply to the liquid state. The ratio $M\lambda/k\eta$ (known as the Prandtl number) for several liquids[45] is given in Table 12.2. It can be seen that near to the triple point the Prandtl number is ~ 2.2, and that it increases with temperature, but is reasonably independent of the liquid chosen. For a dilute gas, this ratio should be 1.5, but C_V/k for the liquids in Table 12.1 is ~ 2.5 (neglecting the internal molecular modes).

The thermal conductivity of a crystal lattice is related to the scattering of high-frequency phonons. Since the thermal conductivity (Table 1.1) of a liquid is similar to that of a solid, it is probable that it is due to similar phenomena. The results of Table 12.2 suggest that the viscosity of a liquid may also be related to the "scattering" of high-frequency modes of motion.

TABLE 12.2
Prandtl number for several liquids

Liquid	Temperature (°K)	$M\lambda/k\eta$
Argon	84·2	2·16
	87·3	2·31
Nitrogen	69·1	2·20
	77·3	2·98
Methane	93·2	2·21
	108·2	2·98
Carbon tetrachloride	298·2	2·19
	318·2	2·78

It is usual to divide the viscosity and the thermal conductivity into kinetic and molecular interaction parts. The former is due to single-particle modes, and for dense liquids is negligible[44,45] compared to the latter component, which is due to co-operative modes. Many authors have considered methods of calculating η and λ. For example Rice and Kirkwood[49] make an approximate calculation of both the viscosity and thermal conductivity coefficients using the following assumptions; (i) the gradient of the potential energy may be expanded as a Taylor series, and only the first two terms need be retained; (ii) the equal time distribution function of momentum and position of a pair of atoms is equal to $g(r)$ times a Maxwellian velocity distribution and a first-order perturbation term; and (iii) the difference in the mean velocities of molecules 1 and 2 at $\mathbf{r}_1$ and $\mathbf{r}_2$ can be Taylor-expanded and only the first two terms retained. With these assumptions they find that the molecular interaction parts of the transport coefficients are—

Shear viscosity (η)

$$= \frac{\rho^2}{30\gamma} \int_V r_2 \left[\frac{\partial^2 u(r)}{\partial r^2} + \frac{4}{r} \frac{\partial u(r)}{\partial r} \right] g(r) \, d\mathbf{r} \qquad (12.21a)$$

Bulk viscosity (ζ)

$$= \frac{\rho^2}{18\gamma} \int_V r^2 \left[\frac{\partial^2 u(r)}{\partial r^2} + \frac{1}{r} \frac{\partial u(r)}{\partial r} \right] g(r) \, d\mathbf{r} \qquad (12.21b)$$

Thermal conductivity (λ)

$$= -\frac{kT}{12\gamma M} \frac{\partial}{\partial T} \left[\rho^2 \int_V r^2 \nabla^2 u(r) g(r) \, d\mathbf{r} \right] \qquad (12.21c)$$

where γ is a friction constant for viscous movements (usually put equal to γ—equation 10.21). From these expressions, the Prandtl number for argon was calculated,[49] and for the two temperatures 84·2 and 87·3°K (Table 12.1) was found to be 2·26 and 2·38, respectively, in fair agreement with experiment. However, over a larger temperature range the agreement is less satisfactory, and the direct calculation[44,45] of η and λ gives worse agreement than the calculation of the ratio. Some comments on equations (12.21) will be made in Chapter 14.

The above experimental and theoretical results suggest that the viscosity and thermal conductivity are related to the high-frequency modes of motion, unlike the diffusion constant which is related to the low-frequency modes. In contrast, Table 12.1 suggests that viscosity is related to the single-particle motion. The variety of viewpoints outlined above illustrates the difficulty inherent in phenomenological analysis, that is an overemphasis of one feature at the expense of another. This difficulty may be overcome only by an adequate basic theoretical treatment.

Symbols for Chapter 12

a	Mean spacing between atoms in liquid	p_j	Probability of atom jumping from one site to another
k_f	Frequency of atom jumping in forward direction	p_V	Probability of atom finding volume to move from one site to another
k_r	Frequency of atom jumping in reverse direction	V_f	Free volume in liquid
k_0	Frequency of atom jumping in absence of force	W	Activation energy
$K(\tau)$	Force in x direction at time τ	γ_s	"Soft" part of friction constant
p_E	Probability of atom finding energy to break bonds to neighbours	γ_h	"Hard" part of friction constant

CHAPTER 13

Co-operative Modes of Motion at Low Frequencies

The thermal motion in a liquid may be divided conveniently into single-particle and co-operative modes of motion, and the next two Chapters will be devoted to a discussion of the latter modes. This subject is a complex one and is not fully understood as yet. The treatment given here is sub-divided into the macroscopic modes, which may be described by hydrodynamic theory, and the microscopic modes, for which the theory is incomplete. A part of this discussion concerns the processes, viscosity and thermal conduction, by which these modes are damped. The viscosity coefficient will be discussed at length, but, because the theory of the thermal conductivity is less well developed and more complex, it will be treated superficially. In the hydrodynamic theory these coefficients are taken as given constants from which the magnitude of the mode damping coefficients may be calculated. Microscopic definitions of the transport coefficients are given and will be generalized for discussion of the microscopic modes of motion.

In this Chapter the theory of sound-wave propagation in a liquid will be described. Although there is an extensive literature on this subject, convenient references are not available for the derivation of some important results. Several of these derivations have been brought together and are given in an Appendix to this Chapter. The basic ideas behind sound-wave theory are discussed in Section 13.2, and a more complete treatment through the visco-elastic theory of liquids is given in Sections 13.3 to 13.5. Finally, the definitions and values of the elastic moduli for a liquid are examined, and the Chapter is concluded with a discussion of the velocity of sound waves.

13.1. The Continuum Model

A co-operative mode of motion is a mode in which more than one atom takes part, and it is distinguished from diffusion processes by the fact that more than one atom is involved in any measurement of these modes. The simplest example is that of long wavelength sound waves: since the wave number is small, the liquid may be treated as a continuum (Section 9.6) and the equations of motion for a continuum employed. Throughout this Chapter the continuum view will be adopted, and the usual macroscopic properties—viscosity, thermal conductivity, etc.—of hydrodynamic theory will be assumed to be relevant. Wave propagation in such a continuum may be divided into longitudinal and transverse components. If $\mathbf{v}$ is the velocity of an element of the continuum, then by definition—

$$\left.\begin{array}{ll} \mathbf{\nabla} \times \mathbf{v} = 0 & \text{for longitudinal waves} \\[2mm] \mathbf{\nabla} \cdot \mathbf{v} = 0 & \text{for tranverse waves} \end{array}\right\} \qquad (13.1)$$

But the continuity equation for the medium is (in linearized form)—

$$\frac{\partial \rho}{\partial t} + \rho \mathbf{\nabla} \cdot \mathbf{v} = 0 \qquad (13.2)$$

so that (for small displacements) longitudinal waves are related to the density fluctuations of the medium, whereas transverse waves are independent of the density fluctuations.

Statistical mechanics shows that the density fluctuations are related to the pressure and entropy fluctuations, which by virtue of the relation $\langle \Delta P \Delta S \rangle = 0$ are independent (equation 1.12). By the method of Chapter 1, starting from equation (1.9), it can be shown that—

$$\langle \Delta V . \Delta P \rangle = - V \chi_s \overline{(\Delta P)^2} = - kT \qquad (13.3a)$$

and—

$$\langle \Delta V . \Delta S \rangle = \left(\frac{\partial T}{\partial P}\right)_s \overline{(\Delta S)^2}$$

$$= kC_P \left(\frac{\partial T}{\partial P}\right)_s$$

$$= k\{(C_P - C_V) TV \chi_T\}^{\frac{1}{2}} \qquad (13.3b)$$

The last step involves the two relationships[1]—

$$\left(\frac{\partial T}{\partial P}\right)_s = \frac{T}{C_P}\left(\frac{\partial V}{\partial T}\right)_P \quad \text{and} \quad (C_P - C_V)\chi_T = \frac{T}{V}\left(\frac{\partial V}{\partial T}\right)_P^2 \qquad (13.3c)$$

Since the density fluctuations are proportional to the volume fluctuations, equations (13.3) will give their magnitude. The pressure fluctuations give a contribution (equations 13.3a and 1.12)—

$$\frac{\langle \Delta V \Delta P \rangle_S^2}{\langle (\Delta P)^2 \rangle} = kTV\chi_S = kTV\chi_T \frac{C_V}{C_P} \tag{13.4a}$$

and the entropy fluctuations give a contribution (equations 13.3b and 1.12)—

$$\frac{\langle \Delta V \Delta S \rangle_P^2}{\langle (\Delta S)^2 \rangle} = \frac{C_P - C_V}{C_P} kTV\chi_T \tag{13.4b}$$

and the sum is—

$$\overline{(\Delta V)^2} = kTV\chi_T \tag{13.4c}$$

as might be expected from equation (1.14) or equation (9.27). Thus the amplitude of thermally excited longitundinal waves can be estimated from equations (13.4), and these modes may be expected to divide into two types given by equations (13.4a) and (13.4b).

13.2. Summary of Sound-wave Propagation

The analysis of sound-wave propagation is based on the idea of writing all the equations of motion to the first order in small quantities, i.e. the equations are linearized. This method may be illustrated in the following way, which also brings out the physical ideas behind the procedure.

The relation between pressure and volume can be written—

$$\Delta P = -B_0 \frac{\Delta V}{V} = -B_0 s_0 \text{ (say)} \tag{13.5}$$

where B_0 is a bulk modulus. Now there is a time lag between the change in pressure and the adjusting movements of the atoms in the liquid, and the deviation of $\Delta V/V$ from its equilibrium value (s_0) will be written $s_i(t)$. Suppose $s_i(t)$ tends to zero at a rate proportional to itself then—

$$-\frac{ds_i(t)}{dt} = \frac{s_i(t)}{\tau_i} \tag{13.6}$$

where τ_i is a relaxation time (this corresponds to exponential damping of s_i). But s_0 is changing continuously because of the thermal fluctuations that occur in the volume, V, of liquid due to its connection to a heat bath. Thus equation (13.6) is modified to—

$$-\frac{ds_i(t)}{dt} + \frac{ds_0(t)}{dt} = \frac{s_i(t)}{\tau_i} \tag{13.7a}$$

or—

$$A_i s_i = (A_i - 1)s_0 \tag{13.7b}$$

where—

$$A_i \equiv \left(1 + \tau_i \frac{d}{dt}\right)$$

If $\tau_i \to \infty$ then this equation gives $s_0 = s_i + \text{constant}$, as expected. Note how this equation includes viscous damping via equation (13.6) and elastic effects via the s_0 term.

Now, the pressure (equation 13.5) must be rewritten to take account of the above effects (and it will be assumed that there are several independent effects), i.e.—

$$\Delta P = -B_0 s_0 - \sum_i B_i s_i = -\left[B_0 + \sum B_i \frac{A_i - 1}{A_i}\right]s_0(t) \tag{13.8}$$

A longitudinal sound wave of frequency ω will cause both ΔP and s_0 to vary as $e^{-i\omega t}$ and thus equation (13.8) reduces to (if q is the wave number of the sound wave)—

$$\frac{\omega^2}{q^2} = c_0^2 - \sum_i c_i^2 \frac{i\omega\tau_i}{1 - i\omega\tau_i} \tag{13.9}$$

where—

$$B_i/M\rho = c_i^2; \qquad \Delta P = \frac{M\rho\omega^2}{q^2}e^{-i\omega t}\bar{s} \quad \text{and} \quad s_0 = \bar{s}\,e^{-i\omega t}$$

Equation (13.9) is of the same form as the dispersion law derived below, and by comparison with the results of Sections 13.4 and 5, the constants may be assigned the values in Table 13.1 (note; since ρ is the number density, $M\rho$ is equal to the mass density).

The sound waves correspond to pressure fluctuations and, usually, it is found that for sound-wave frequencies $\omega \gg D_T q^2$, where $D_T = \lambda/C_V\rho$. In these circumstances, equation (13.9) reduces to the Navier–Stokes equation, or—

$$\omega^2 \simeq \frac{q^2}{M\rho\chi_s} - \frac{i\omega q^2}{M\rho}(\tfrac{4}{3}\eta + \zeta) \tag{13.10}$$

TABLE 13.1

Constants in sound-wave formula

Effect	c_i^2	τ_i
Shear Viscosity (η)	$\dfrac{4}{3}\dfrac{G}{M\rho}$	$\dfrac{\eta}{G}$
Bulk Viscosity (ζ)	$\dfrac{B_1}{M\rho}$	$\dfrac{\zeta}{B_1}$
Thermal Conduction (λ)	$\dfrac{C_P - C_V}{C_V} \cdot \dfrac{1}{M\rho\chi_T}$	$\dfrac{\rho C_V}{\lambda q^2}$

G is the shear modulus; η the shear viscosity and ζ the bulk viscosity.

At the sound-wave frequency of—

$$\omega \sim q\sqrt{\frac{1}{M\rho\chi_s}} \quad \text{a term} \quad \left[\frac{i\omega q^2 \lambda}{M\rho}\left(\frac{1}{C_V} - \frac{1}{C_P}\right)\right]$$

comes from equation (13.9) and is added to equation (13.10), and the two imaginary terms give rise to the damping of sound waves. In the low-frequency region corresponding to $\omega \sim D_T q^2$ equation (13.10) takes on a different form—

$$0 \sim \omega^2 \simeq \frac{q^2}{M\rho\chi_T} - \frac{C_P - C_V}{C_V}\frac{q^2}{M\rho\chi_T}\frac{i\omega}{D_T q^2 - i\omega} \tag{13.11}$$

and in this case there is no propagating term. The entropy fluctuations are described by equation (13.11) and as expected from equation (13.4b) are proportional to $C_P - C_V$.

The basis of this method will be discussed in more detail in Section 13.3, and the theory for transverse and longitudinal waves will be discussed separately in Sections 13.4 and 13.5, respectively.

13.3. Basic Equations of Motion in Visco-elastic Theory

(The discussion of Sections 13.3 and 13.5 and of A1 and A2 in the Appendix is taken from the unpublished notes of W. C. Marshall; a general discussion of visco-elastic theory can be found in Frenkel[51] and Lodge.[52])

In this Section the equations of motion for an element of liquid (moving with velocity **v**) will be obtained. The force acting in the α

direction over a unit area at right angles to the β direction will be denoted by $P_{\alpha\beta}$; Fig. 13.1(a) shows the co-ordinate system. If the thickness of the volume element (of unit area) is Δr_β, then the force $F_{\alpha\beta}$ (per unit area) shown in Fig. 13.1(b); is—

$$F_{\alpha\beta} = P_{\alpha\beta} + \Delta r_\beta \frac{\partial P_{\alpha\beta}}{\partial r_\beta} \qquad (13.12a)$$

Since the net force is $F_{\alpha\beta} - P_{\alpha\beta}$, the equation of motion is—

$$M_1 \frac{\partial v_\alpha}{\partial t} = \Delta r_\beta \frac{\partial P_{\alpha\beta}}{\partial r_\beta} \qquad (13.12b)$$

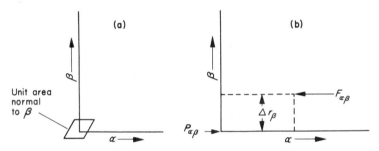

Fig. 13.1. (a) Illustration of co-ordinates for equation (13.12). (b) Forces on volume element of liquid.

where M_1 is the mass (per unit area) of the volume element. The components of force for other directions are added to the right-hand side of this equation to give finally—

$$M\rho \frac{\partial v_\alpha}{\partial t} = \nabla_\beta P_{\alpha\beta} \qquad (13.13)$$

where ρ is the number density and M is the mass of a molecule, (in the usual tensor notation ∇_β includes a sum over β).

In order to obtain an expression for $P_{\alpha\beta}$, it is necessary to make an extension of the usual hydrodynamic model and this will be done through the visco-elastic theory of liquids. In the hydrodynamic model of liquids, it is assumed that there is little or no resistance to shear forces. Although this is true if the forces are applied slowly over a long period of time, it is far from true for high-frequency alternating forces. In the latter case, the molecules of the liquid do not have time to adjust their positions by inelastic (or viscous) movement over the interval during which the force is applied, and the only possible motion is that of elastic deformation analogous to the behaviour of a solid. Thus in general there will be two types of motion, viscous movements in which energy is dissipated and elastic deformations without dissipation of energy. The combination of these two effects into one theory is known as the visco-elastic theory and several variations of this theory have been discussed in detail by Frenkel.[51] In the simplest version it is assumed the viscous and elastic effects may be treated independently and terms corresponding to the two effects added linearly in the equations of motion. Moreover, in the simple theory, each relaxation process is assumed to occur with a single relaxation time as described in Section 13.2. The required equation for $P_{\alpha\beta}$ is derived in the Appendix to this chapter (equation A 13.8), and is—

$$\eta[\nabla_\alpha v_\beta + \nabla_\beta v_\alpha - \tfrac{2}{3}\dot{\nabla}_\gamma v_\gamma \delta_{\alpha\beta}] + \zeta \mathbf{V} \cdot \mathbf{v}\delta_{\alpha\beta}$$

$$= \left\{1 + \tau_m \frac{\partial}{\partial t}\right\}\left\{P_{\beta\alpha} + P\delta_{\alpha\beta}\right\} \tag{13.14}$$

where P is the hydrostatic pressure and $\tau_m = \eta/G$ is the Maxwell relaxation time. The right-hand side of this equation is the sum of the viscous and elastic terms referred to above. Finally, the combination of equations (13.13) and (13.14) gives the equation of motion for the volume element being considered.

13.4. Propagation of Transverse Modes

A simple example of the use of this model is a calculation of the frequency–wave number relationship for transverse waves. In this case, it can be seen that—

$$\alpha \equiv y; \qquad \beta \equiv x$$

$$\mathbf{V} \equiv \left(\frac{d}{dx}, 0, 0\right); \qquad \mathbf{l} \equiv (0, l_0\, e^{i(qx - \omega t)}, 0)$$

$$P_{yx} = -P + P'_{yx}\, e^{i(qx - \omega t)}$$

where ω is the frequency and q the wave-number of the wave. The quantity $\mathbf{l}$ is the displacement and is related to the velocity by equation (A 13.5). Thus equation (13.13) becomes—

$$-M\rho\omega^2 l_0 = iqP'_{yx} \tag{13.15}$$

and equation (13.14) becomes—

$$\left(\frac{1}{\eta} - \frac{i\omega}{G}\right)P'_{yx} = (iq)(-i\omega)l_0 \tag{13.16}$$

Elimination of P'_{yx}/l_0 from equation (13.15) and (13.16) gives—

$$\omega^2 = \frac{1}{M\rho} \frac{-i\omega q^2}{(1/\eta)-(i\omega/G)} = -\frac{q^2 G}{M\rho} \cdot \frac{i\omega\tau_m}{1-i\omega\tau_m} \qquad (13.17)$$

Equation (13.17) will be recognized as the usual expression for the dispersion of transverse waves, involving the Maxwell relaxation time τ_m, and high-frequency velocity $\sqrt{G/M\rho}$. Figure 13.2 is a qualitative graph of equation 13.17: at low frequencies, the transverse modes are not propagated, but for $\omega \gg 1/\tau_m$ propagation is possible as in a solid.

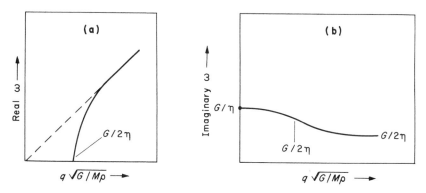

Fig. 13.2. Qualitative behaviour of equation (13.17). (a) Real part of ω. (b) Imaginary part of ω.

Since transverse modes of motion are not propagated at low frequencies, it is not possible to examine their properties by normal ultrasonic techniques. However the scattering of light (in which the plane of polarization is rotated through a right-angle) affords a suitable method. The "transverse mode" scattering is seen as a line whose mean frequency is equal to the frequency of the incident radiation. At low frequencies ($\ll 1/\tau_m$) the width of this line is obtained from equation (13.17) as—

$$\Delta\omega = \frac{q^2 G \tau_m}{M\rho} = q^2 \frac{\eta}{M\rho} \qquad (13.18)$$

The intensity may be calculated[29] in terms of the strain tensor (u), since the intensity of light scattering (Section 8.3) is related to the fluctuations in the dielectric constant, and these are related to the strains in the medium. For transverse motion the fluctuation ($\delta\varepsilon$) of the dielectric constant may be written—

$$\delta\varepsilon_{\alpha\beta} = au_{\alpha\beta} \qquad (13.19)$$

where $\alpha \neq \beta$ and a is coefficient related to the atomic polarizability. If the incident wave-vector is $\mathbf{k}_0$ and the wave vector of the scattered light is $\mathbf{k}$, then it can be shown that scattering of the incident wave (electric field $\mathbf{E}$, at right angles to $\mathbf{k}_0$) occurs only for the component of transverse motion at right angles to the $(\mathbf{k}_0, \mathbf{k})$ plane. In this case Landau and Lifshitz[29] show that the electric field of the scattered wave for an angle of scatter θ, is given by—

$$\left. \begin{aligned} E_{\parallel} &= \frac{\exp ik_0r}{r} \cdot \frac{ik_0^2 aqu_0 V \cos \theta/2}{16\pi} E_{0\perp} \\ E_{\perp} &= \frac{\exp ik_0r}{r} \cdot \frac{ik_0^2 aqu_0 V \cos \theta/2}{16\pi} E_{0\parallel} \end{aligned} \right\} \tag{13.20}$$

where u_0 is the displacement due to the mode and $\mathbf{E}_0$ is the field of the incident wave ($\parallel$ and $\perp$ indicate components of the field lying in and perpendicular to the plane of $\mathbf{k}_0, \mathbf{k}$). Since the coefficient in each expression is the same the cross-section is given by the square of the modulus of the wave amplitude, or—

$$\left(\frac{d\sigma}{d\Omega} \right)_{\text{light}} = \left| \frac{qk_0^2 a}{16\pi} \right|^2 \frac{V\overline{u_0^2}}{\rho} \cos^2\theta/2 \tag{13.21}$$

(note how the angular factor in this equation differs from that in equation 8.17b). The mean square amplitude $\overline{u_0^2}$ of the mode is equal to the same quantity[29] in the absence of viscous damping, or—

$$\overline{u_0^2} = \frac{4kT}{VGq^2} \tag{13.22}$$

Finally equations (13.21) and (13.22) give the scattered intensity in terms of the coefficient a (equation 13.19).

This method of studying transverse modes has rarely been used for liquids because usually the value of a is quite small, but it is commonly employed in the case of amorphous solids.

13.5. Dispersion of Longitudinal Modes

In the case of longitudinal waves, the quantities in equations (13.13 and 13.14) become—

$$\left. \begin{aligned} \alpha &\equiv x; \quad \beta \equiv x; \\ \mathbf{l} &\equiv (l_0 \, e^{iqx - i\omega t}, 0, 0) \\ P &= P_0 + \Delta P \, e^{iqx - i\omega t} \\ P_{xx} &= -P_0 + P'_{xx} \, e^{iqx - i\omega t} \end{aligned} \right\} \tag{13.23}$$

The quantity P_0 is the uniform component of the hydrostatic pressure, and ΔP is a pressure fluctuation (thermally excited) which propagates as a longitudinal wave. Equation (13.13) is now—

$$-M\rho\omega^2 l_0 = iqP'_{xx} \tag{13.24}$$

and equation (13.14) becomes—

$$\left\{1 - \frac{i\omega\eta}{G}\right\}\{P'_{xx} + \Delta P\} = (\tfrac{4}{3}\eta + \zeta)q\omega l_0 \tag{13.25}$$

Thus this problem is more complex than the transverse case because the fluctuation of the hydrostatic pressure (ΔP) must be related to P'_{xx} in order to reduce equation (13.25) to a useful form. Since P'_{xx} is proportional to l_0 (equation (13.24) it is assumed that ΔP behaves in a similar way, i.e.—

$$\Delta P = -iqCl_0 \tag{13.26}$$

and then equation (13.25) reduces to—

$$\omega^2 = \frac{q^2 C}{M\rho} - \frac{iq^2\omega}{M\rho} \frac{\tfrac{4}{3}\eta + \zeta}{1 - i\omega\tau_m} \tag{13.27}$$

so that C is equivalent to an elastic modulus.

The coefficient C is determined by noting that—

$$\chi_T \Delta P = \frac{\Delta\rho}{\rho} + \left(\frac{\partial V}{\partial T}\right)_P \frac{\Delta T}{V} \tag{13.28}$$

and that by integrating the continuity equation (13.2)—

$$\frac{\Delta\rho}{\rho} = \nabla \cdot \mathbf{l} = -iql_0 \tag{13.29}$$

For isothermal processes, the ΔT term is neglected and then equations (13.26) and (13.28) show that $C = 1/\chi_T$. In this case, the dispersion curve has the form shown at Fig. 13.3(a).

The temperature fluctuation can be deduced from the energy equation as shown in the Appendix (A2). In this case, equation (A 13.15) reduces equations (13.28) and (13.26) to—

$$C = \frac{1}{\chi_T}\left[1 - \frac{C_P - C_V}{C_V} \frac{i\omega}{\omega_0 - i\omega}\right] \tag{13.30}$$

where—

$$\omega_0 = \frac{\lambda q^2}{\rho C_V} = D_T q^2$$

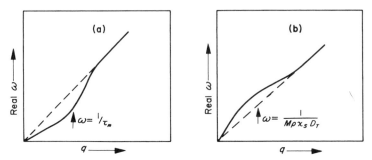

FIG. 13.3. Qualitative behaviour of equation (13.27). (a) Real part of ω for viscous terms only. (b) Real part of ω for heat conduction terms only.

Thus the ΔT term in equation (13.28) leads to a term proportional to $C_P - C_V$ as might have been expected from the discussion of Section 13.1. The complete expression for sound wave propagation is obtained by substituting equation (13.30) into equation (13.27), and comparison of this equation with equation (13.9) leads to the results quoted in Table 13.1. If the viscous term in equation (13.27) is negligible, the frequency dependence of the sound velocity will be given by equation (13.30). In this case the sound velocity will be proportional $\sqrt{1/\chi_S}$ at low frequencies and $\sqrt{1/\chi_T}$ at high frequencies; Figure 13.3(b) shows this behaviour qualitatively which is opposite in effect to that shown in Fig. 13.3(a).

For simple liquids, such as argon or sodium, the region of high dispersion (shown in Fig. 13.3) may be calculated to occur at $q \sim 0.25$ $Å^{-1}$. At such (relatively) high wave numbers, the continuum theory is no longer valid, and hence the above treatment can be valid only over the low q portion of Fig. 13.3. The relaxation time for simple liquids is $\sim 10^{-12}$ sec, but in complex liquids relaxation times are longer and are often in the region of 10^{-7} sec. In this case, the above theory should be valid and ultrasonic techniques can be used to obtain experimental data. An example is given in Fig. 13.4 where data on zinc chloride are compared to a single term in equation (13.9): experimental frequencies between 5 and 45×10^6 cycles/sec were used and the values of τ, c_0 and c_1 were obtained by fitting to the data. Zinc chloride contains complex ions, and it is believed that the relaxation process involves the breakdown of the ionic structure; it is an interesting case since only one term in equation (13.9) is required. Ultrasonic data usually require several terms in the sum in equation (13.9) to give an adequate theoretical fit.

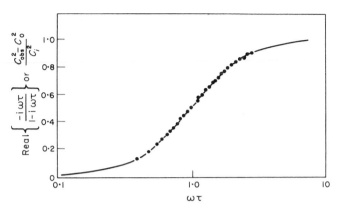

FIG. 13.4. Real component of the square of the sound velocity as given by equation (13.9). The full line is the theoretical curve and the circles are the experimental values for molten zinc chloride (from Grubert, G. J., and Litovitz, T. A., *J. Chem. Phys.* 1964, **40**, 13, Fig. 6). Data covering temperatures between 320° and 400°C were used, and reduced to a common curve the the Arrhenius formula (Section 12.2).

13.6. Elastic Moduli

In some cases it is convenient to discuss fluctuation problems in Fourier space (i.e. Q space) and obtain macroscopic properties by taking the $Q \to 0$ limit. A discussion of the elastic moduli is an example because the stress tensor can be written out most conveniently in Q space. The principle of this method is that a small volume of the liquid is defined by the wavelength $2\pi/Q$, and the thermal fluctuations in this volume are taken into account by taking the thermal average. After this step the macroscopic property, which is a function of the microscopic fluctuations, is obtained as the $Q \to 0$ limit. Schofield,[23] in a fundamental paper, obtains expressions for the elastic moduli: his method is given below and in Appendix A3 and A4.

First the atomic current density ($\mathbf{j}$) is written in terms of the velocity $v_{i\alpha}$ (the α component of the velocity of the ith atom)—

$$j_\alpha = \sum_{i=1}^{N} v_{i\alpha}\delta(\mathbf{r}-\mathbf{r}_i) \tag{13.31}$$

which on Fourier-transforming becomes—

$$(j_\alpha)_Q = \sum_{i=1}^{N} v_{i\alpha}\,e^{i\mathbf{Q}\cdot\mathbf{r}_i} = \sum_{i=1}^{N} v_{i\alpha}\,e^{iQ_\beta r_{i\beta}} \tag{13.32}$$

where the subscript Q denotes the Qth Fourier component and the suffix β indicates the $\mathbf{Q}$ direction. Secondly, the momentum conservation equation is used to define the stress tensor (this equation is equivalent to equation 13.13)—

$$M \frac{\partial j_\alpha}{\partial t} = -\nabla_\beta \sigma_{\alpha\beta} \tag{13.33}$$

On Fourier-transforming this equation it becomes—

$$M \left(\frac{\partial j_\alpha}{\partial t} \right)_Q = iQ_\beta (\sigma_{\alpha\beta})_Q \tag{13.34}$$

Equation (13.32) may be substituted into the left-hand side of this equation in order to express $(\sigma_{\alpha\beta})_Q$ in microscopic quantities. This step is carried through in the Appendix A3 and the $Q = 0$ value is—

$$\{\sigma_{\alpha\beta}\}_0 = \sum_{i=1}^{N} \left\{ M v_{i\alpha} v_{i\beta} - \frac{1}{2} \sum_{j \neq i} \frac{r_{ij\alpha} r_{ij\beta}}{r_{ij}} \frac{du(r_{ij})}{dr_{ij}} \right\} \tag{13.35}$$

where $u(r)$ is the pair potential and (i, j) denote the ith and jth atoms. This definition together with equation (2.24) leads immediately to the interesting result $\langle \{\sigma_{\alpha\alpha}\}_0 \rangle = VP$.

Now the elastic moduli may be defined as the ratio of the rate of change of the stress tensor to the rate of strain tensor—

$$\left(\frac{\partial \sigma_{\alpha\beta}}{\partial t} \right)_Q = \frac{1}{\rho} (c_{\alpha\beta\gamma\delta})_Q (\Lambda_{\gamma\delta})_Q + \text{terms that average to zero} \tag{13.36}$$

where $c_{\alpha\beta\gamma\delta}$ is an elastic modulus and $\Lambda_{\gamma\delta}$ is a component of the rate of strain tensor. Also the rate of strain (defined in equation A 13.1) may be expressed simply as (equation A 13.24)—

$$(\Lambda_{\alpha\beta})_Q = \frac{1}{2} i [Q_\alpha (j_\beta)_Q + Q_\beta (j_\alpha)_Q] \tag{13.37}$$

Thus equation (13.36), when used with equations (13.35) and (13.37) gives a microscopic definition of the elastic moduli. The details of this step are given in the Appendix A4.

For a symmetrical system there are three independent elastic constants which are given in equation (A 13.27). A simple way of writing these equations is to express them in terms of the following integrals—

$$\left. \begin{aligned} I_1 &= \frac{\rho}{2kT} \int_V g(r) r \frac{du(r)}{dr} \, d\mathbf{r} \\ I_2 &= \frac{\rho}{2kT} \int_V g(r) r^2 \frac{d^2 u(r)}{dr^2} \, d\mathbf{r} \end{aligned} \right\} \tag{13.38}$$

Then equations (A 13.27) become (using the usual Voigt notation for the elastic constants)—

$$c_{11} = \rho k T(3 + \tfrac{2}{5}I_1 + \tfrac{1}{5}I_2)$$
$$c_{12} = \rho k T(1 - \tfrac{2}{5}I_1 + \tfrac{1}{15}I_2)$$
$$c_{44} = \rho k T(1 + \tfrac{4}{15}I_1 + \tfrac{1}{15}I_2)$$

(13.39)

Because the existence of central forces has been assumed, these expressions satisfy the Cauchy relationship $c_{11} - c_{12} = 2c_{44}$, which reduces the number of independent constants to two. The integral I_1 will have been recognized as the integral given at equation (7.10) and takes on the value 3 at the triple point. It will be shown in the next Section that the integral I_2 is > 300, so that the elastic constants have the following approximate values—

$$c_{44} \simeq c_{12} \simeq c_{11}/3 \simeq \rho k T I_2/15$$

(13.40)

at the triple point. The shear modulus, G, is equal to C_{44}, and is given in terms of the other moduli by this equation.

13.7. The Velocity of Sound

The velocity of sound is equal to the square root of the product of the density and a modulus (equation 13.9, for example). It is interesting to compare the values of the moduli that appear in such an expression for varying conditions. At low frequency, for example, the relevant modulus is $1/\chi_S$ but at the high-frequency limit of equation (13.9), it is $1/\chi_T + B_1 + \tfrac{4}{3}G$. A useful result is obtained by considering the harmonic solid-like limit to equation (13.9). In this case, the frequency is low compared to $1/\tau_i$ for thermal conduction, but high for the bulk viscosity term, and the bulk modulus (B) then becomes $B = 1/\chi_S + B_1$. It is a well known result of solid-state theory that for an isotropic solid $B = \tfrac{1}{3}(c_{11} + 2c_{12})$ and thus—

$$\frac{1}{\chi_S} \leq B = \frac{\rho k T}{3}[5 - \tfrac{2}{3}I_1 + \tfrac{1}{3}I_2]$$

(13.41)

But since $\chi_T \geq \chi_S$ this equation may be extended (via equation 6.23) to—

$$\frac{1}{S(0)} \leq \frac{1}{\rho k T \chi_S} \leq \{\tfrac{5}{3} - \tfrac{2}{9}I_1\} + \tfrac{1}{9}I_2$$

(13.42)

As discussed in Chapter 7, a typical value of $S(0)$ at the triple point is

~ 0.03 and $I_1 \sim 3$, so that (13.42) becomes—

$$I_2 \gtrsim 300 \text{ (at the triple point)} \tag{13.43}$$

This result emphasizes that for low compressibility liquids the value of I_2 is very large and gives the major contribution to the velocity of sound.

Another interesting case is that of a liquid metal, where an approximate value of the velocity of sound may be calculated[13] on the basis of plasma theory in the limit of small Z. If it is assumed that the metal ions may be treated as a plasma, the velocity of sound is equal to the plasma frequency (Ω). However the ions are screened by the conduction electrons and, as in Chapter 4, this screening may be included by dividing by the dielectric function, $\varepsilon(q)$; thus—

$$c^2 = \frac{\Omega^2}{q^2 \varepsilon(q)} \simeq \frac{2}{3} \cdot \frac{Z^2 E_f}{M} \quad \text{for small } Z \text{ and } q \to 0 \tag{13.44}$$

In principle, this equation could serve as an internal consistency check upon $u(r)$ for a metal, since $u(r)$ is calculated from E_f (and thus I_2 may be calculated from E_f) but this has not yet been done.

Appendix

A1. Derivation of the Pressure Tensor in Visco-elastic Theory

The internal force acting on the volume element of liquid is the sum of several terms. First, there is the hydrostatic pressure (P), which is dependent on the instantaneous positions of the atoms and hence contributes a term $-P\delta_{\alpha\beta}$ to the force $P_{\beta\alpha}$; secondly, the viscous forces are related to the rate of strain tensor Λ, the components of which are defined (in a form symmetrical in α, β) as—

$$\Lambda_{\alpha\beta} = \tfrac{1}{2}(\nabla_\alpha v_\beta + \nabla_\beta v_\alpha) \tag{A 13.1}$$

Hence from the definition of viscosity (as the ratio of force to rate of strain), it is seen that—

$$P_{\beta\alpha} + P\delta_{\alpha\beta} = 2\eta \Lambda''_{\alpha\beta} + \eta' \Lambda''_{\gamma\gamma}\delta_{\alpha\beta} \tag{A 13.2}$$

where the double prime indicates the strain component due to viscous effects, η is the shear viscosity and η' the dilational viscosity. It is conventional[10] to rewrite equation (A 13.2) in terms of a bulk viscosity (ζ), defined as—

$$\zeta = \eta' + \tfrac{2}{3}\eta \tag{A 13.3}$$

so that—

$$P_{\beta\alpha} + P\delta_{\alpha\beta} = 2\eta[\Lambda''_{\alpha\beta} - \tfrac{1}{3}\Lambda''_{\gamma\gamma}\delta_{\alpha\beta}] + \zeta\,\mathbf{V}\,.\,\mathbf{v}''\delta_{\alpha\beta} \qquad (A\ 13.4)$$

The reason for this division is that the term in equation (A 13.4) proportional to η gives the effect of changing shape without changing volume, and the term proportional to ζ gives the effect of internal friction due to the change of volume. In an incompressible fluid $\mathbf{V}\,.\,\mathbf{v} = 0$ and the latter term vanishes. These two terms will be treated independently through the assumption that the strains due to changes in shape or volume are independent effects.

The elastic terms depend upon the instantaneous strains, which are written in terms of a displacement, $\mathbf{l}$, defined by—

$$\mathbf{v} = \frac{\partial \mathbf{l}}{\partial t} \qquad (A\ 13.5)$$

Thus the elastic terms are—

$$P_{\beta\alpha} + P\delta_{\alpha\beta} = G\{\nabla_\alpha l_\beta + \nabla_\beta l_\alpha - \tfrac{2}{3}\mathbf{V}\,.\,\mathbf{l}\delta_{\alpha\beta}\} + B_1\mathbf{V}\,.\,\mathbf{l}\delta_{\alpha\beta} \qquad (A\ 13.6)$$

where G is the shear modulus (see Section 13.6) and B_1 is a "relaxation" bulk modulus. Strains corresponding to this term decay with a lifetime related to the bulk viscosity.[51] Equation (A 13.6) may be rewritten (where the prime indicates strain due to elastic effects) as—

$$\frac{\partial}{\partial t}\,.\,\frac{1}{G}\{P_{\beta\alpha} + P\delta_{\alpha\beta}\} = 2[\Lambda'_{\alpha\beta} - \tfrac{1}{3}\Lambda'_{\gamma\gamma}\delta_{\alpha\beta}] + \frac{B_1}{G}\mathbf{V}\,.\,\mathbf{v}'\delta_{\alpha\beta} \qquad (A\ 13.7)$$

Now the total rate of strain is given by the sum of the left-hand sides of (A 13.4) and (A 13.7), i.e.—

$$\left\{\frac{1}{\eta} + \frac{1}{G}\frac{\partial}{\partial t}\right\}\{P_{\beta\alpha} + P\delta_{\alpha\beta}\} = 2[\Lambda_{\alpha\beta} - \tfrac{1}{3}\Lambda_{\gamma\gamma}\delta_{\alpha\beta}] + A\mathbf{V}\,.\,\mathbf{v}\delta_{\alpha\beta} \qquad (A\ 13.8)$$

In writing (A 13.8) the equal relaxation time hypothesis has been adopted, i.e.—

or—

$$\left.\begin{aligned}\tau_m &= \frac{\eta}{G} = \frac{\zeta}{B_1} \\[2mm] A &= \frac{\zeta}{\eta} = \frac{B_1}{G}\end{aligned}\right\} \qquad (A\ 13.9)$$

Equation (A 13.8) is the final equation for the components of the pressure tensor, and is symmetrical in the suffixes α, β.

A2. Derivation of Frequency-dependent Thermal Expansion

The starting point for this derivation is the equation expressing the rate of change of energy in the volume element as the sum of the thermal flux and the change in the strain energy, i.e.—

$$\rho \frac{dE}{dt} = -\mathbf{V} \cdot \mathbf{q} + \Lambda_{\alpha\beta} P_{\alpha\beta} \qquad (A\ 13.10)$$

where E is the internal energy, and $\mathbf{q}$ is the thermal flux, which is related to the thermal conductivity (λ) by $\mathbf{q} = -\lambda \mathbf{V} T$. But since the energy is a function of temperature and volume—

$$\frac{dE}{dt} = \frac{dT}{dt}\left(\frac{\partial E}{\partial T}\right)_V + \frac{dV}{dt}\left(\frac{\partial E}{\partial V}\right)_T$$

$$= \frac{dT}{dt}C_V + \frac{1}{\rho^2}\frac{\partial \rho}{\partial t}\left[NT\left(\frac{\partial V}{\partial T}\right)_P \left(\frac{\partial P}{\partial V}\right)_T + P\right] \qquad (A\ 13.11)$$

Then by using equation (13.2) in (A 13.11) and equating the result to equation (A 13.10), the following equation is obtained by use of equation (A 13.1)—

$$\rho C_V \frac{dT}{dt} = \lambda \mathbf{V}^2 T - \frac{NT}{V}\left(\frac{\partial V}{\partial T}\right)_P \frac{1}{\chi_T} \mathbf{V} \cdot \mathbf{v} + \Lambda_{\alpha\beta}(P_{\alpha\beta} + P\delta_{\alpha\beta}) \quad (A\ 13.12)$$

It is seen from equation (13.23) that for a sound wave, the pressure term in this equation vanishes to first order. Thus with the substitution—

$$T = T_0 + \Delta T\, e^{i(qx - \omega t)} \qquad (A\ 13.13)$$

the following result is obtained from equations (13.23) and (A 3.12)—

$$-i\omega\rho C_V \Delta T = -\lambda q^2 \Delta T - q\omega l_0 \frac{T}{V}\left(\frac{\partial V}{\partial T}\right)_P \frac{N}{\chi_T} \qquad (A\ 13.14)$$

Finally equation (13.3c) is used in equation (A 13.14) to show that—

$$\Delta T\left(\frac{\partial V}{\partial T}\right)_P = -\frac{Nq\omega l_0(C_P - C_V)}{\lambda q^2 - i\omega\rho C_V}$$

This equation is the desired result, giving the frequency-dependent volume change corresponding to a temperature change ΔT.

A3. Derivation of the Microscopic Definition of the Stress in a Liquid

Equation (13.34) of the text is used as the starting point, i.e.—

$$M\left(\frac{\partial j_\alpha}{\partial t}\right)_Q = iQ_\beta(\sigma_{\alpha\beta})_Q \qquad \text{(A 13.16)}$$

The left-hand side of this equation is evaluated from equation (13.32)—

$$\left(\frac{\partial j_\alpha}{\partial t}\right)_Q = iQ_\beta \sum_{i=1}^{N} \left\{ v_{i\alpha} v_{i\beta} - \frac{1}{iQ_\beta M} \sum_{j \neq i} \frac{du(r_{ij})}{dr_{i\alpha}} \right\} e^{iQ \cdot r_i} \qquad \text{(A 13.17)}$$

where $u(r)$ is the pair potential and $r_{ij} = r_j - r_i$. The second term in the curly brackets of this equation can be rewritten as (note the minus sign is included)—

$$\frac{1}{iQ_\beta M} \sum_{j \neq i} \frac{r_{ij\alpha}}{|r_{ij}|} \frac{du(r_{ij})}{dr_{ij}} e^{iQ \cdot r_i} = \frac{1}{iQ_\beta M} \sum_{j \neq i} \frac{r_{ji\alpha}}{|r_{ji}|} \frac{du(r_{ji})}{dr_{ji}} e^{iQ \cdot r_i} \qquad \text{(A 13.18)}$$

But $r_{ji\alpha} = -r_{ij\alpha}$, and each term in equation (A 13.18) is equal to half the sum of the two terms, so that each can be written—

$$= -\frac{1}{2iQ_\beta M} \sum_{j \neq i} \frac{r_{ij\alpha}}{|r_{ij}|} \frac{du(r_{ij})}{dr_{ij}} [e^{iQ \cdot r_j} - e^{iQ \cdot r_i}]$$

$$= -\frac{1}{2M} \sum_{j \neq i} \frac{r_{ij\alpha} r_{ij\beta}}{r_{ij}} \frac{du(r_{ij})}{dr_{ij}} \frac{e^{iQ \cdot r_{ij}} - 1}{iQ \cdot r_{ij}} e^{iQ \cdot r_i} \qquad \text{(A 13.19)}$$

The latter result has the advantage of being symmetrical in the components (α, β) and (ij). Thus combining equations (A 13.19) and (A 13.17) with (A 13.16) the Qth component of the stress tensor is given by—

$$(\sigma_{\alpha\beta})_Q = \sum_{i=1}^{N} \left\{ M v_{i\alpha} v_{i\beta} - \frac{1}{2} \sum_{j \neq i} \frac{r_{ij\alpha} r_{ij\beta}}{r_{ij}} \frac{du(r_{ij})}{dr_{ij}} \frac{e^{iQ \cdot r_{ij}} - 1}{iQ \cdot r_{ij}} \right\} e^{iQ \cdot r_i} \qquad \text{(A 13.20)}$$

and the $Q = 0$ limit gives the macroscopic stress tensor.

A4. Derivation of the Microscopic Definitions of the Elastic Moduli

The elastic moduli may be defined in terms of the ratio of the rate of change of the stress tensor to the rate of strain tensor (equation 13.36), i.e.—

$$\left(\frac{\partial \sigma_{\alpha\beta}}{\partial t}\right)_Q = \frac{1}{\rho} (c_{\alpha\beta\gamma\delta})_Q (\Lambda_{\gamma\delta})_Q + \left(\frac{\partial \sigma'_{\alpha\beta}}{\partial t}\right)_Q \qquad \text{(A 13.21)}$$

where $c_{\alpha\beta\gamma\delta}$ is an elastic modulus and $\partial\sigma'/\partial t$ is orthogonal to $\Lambda_{\alpha\beta}$.

Multiplication of this equation by $(\Lambda_{\lambda\mu})_{-Q}$ and thermal averaging gives—

$$\left\langle (\Lambda_{\lambda\mu})_{-Q}\left(\frac{\partial\sigma_{\alpha\beta}}{\partial t}\right)_Q \right\rangle = \frac{1}{\rho}(c_{\alpha\beta\gamma\delta})_Q \langle (\Lambda_{\lambda\mu})_{-Q}(\Lambda_{\gamma\delta})_Q \rangle \qquad \text{(A 13.22)}$$

It should be noted that the $(-Q, +Q)$ multiplication arises from the fact that a physical quantity is proportional to $\langle \phi^*\phi \rangle$. The stationarity condition gives—

$$\left\langle (\Lambda_{\lambda\mu})_{-Q}\left(\frac{\partial\sigma_{\alpha\beta}}{\partial t}\right)_Q \right\rangle = -\left\langle \left(\frac{\partial\Lambda_{\lambda\mu}}{\partial t}\right)_{-Q}(\sigma_{\alpha\beta})_Q \right\rangle \qquad \text{(A 13.23)}$$

To reduce this equation the Fourier transform of equation (13.36) is taken, i.e.—

$$(\Lambda_{\alpha\beta})_Q = \tfrac{1}{2}i[Q_\alpha(j_\beta)_Q + Q_\beta(j_\alpha)_Q] \qquad \text{(A 13.24)}$$

and equation (11.24b) is rewritten as—

$$\langle (j_\alpha)_{-Q}(j_\beta)_Q \rangle = N\frac{kT}{M}\delta_{\alpha\beta} \qquad \text{(A 13.25)}$$

Then equations (A 13.24) and (A 13.16) are substituted in the right-hand side of equation (A 13.23), and (A 13.24) followed by equation (A 13.25) are substituted into the right-hand side of equation (A 13.22). In this case, equations (A 13.22) and (A 13.23) reduce to the basic equation—

$$\tfrac{1}{2}\langle \{Q_\lambda Q_\nu(\sigma_{\mu\nu})_{-Q} + Q_\mu Q_\nu(\sigma_{\lambda\nu})_{-Q}\}(\sigma_{\alpha\beta})_Q \rangle$$
$$= \frac{V\rho kT}{4}(c_{\alpha\beta\gamma\delta})_Q\{Q_\lambda Q_\gamma\delta_{\lambda\gamma} + Q_\mu Q_\gamma\delta_{\mu\gamma} + Q_\lambda Q_\delta\delta_{\lambda\delta} + Q_\mu Q_\delta\delta_{\mu\delta}\}$$

$$\text{(A 13.26)}$$

A symmetrical system such as a liquid has only three independent[10] elastic constants which from (A 13.26) are given in the $Q \to 0$ limit by (if the z axis is in the $\mathbf{Q}$ direction)—

$$c_{11} = \frac{1}{VkT}\langle \{\sigma'_{zz} \cdot \sigma'_{zz}\}_{Q\to0} \rangle = 3\rho kT + \frac{\rho^2}{2}\int_V g(r)\frac{\partial^2 u(r)}{\partial z^2}z^2\,d\mathbf{r} \quad \text{(A 13.27a)}$$

$$c_{12} = \frac{1}{VkT}\langle \{\sigma'_{zz} \cdot \sigma'_{xx}\}_{Q\to0} \rangle = \rho kT + \frac{\rho^2}{2}\int_V g(r)\left[xz\frac{\partial^2 u(r)}{\partial x\partial z} - x\frac{\partial u(r)}{\partial x}\right]d\mathbf{r}$$

$$\text{(A 13.27b)}$$

$$c_{44} = \frac{1}{VkT}\langle \{\sigma'_{zx} \cdot \sigma'_{zx}\}_{Q\to0} \rangle = \rho kT + \frac{\rho^2}{2}\int_V g(r)\frac{\partial^2 u(r)}{\partial x^2}z^2\,d\mathbf{r} \quad \text{(A 13.27c)}$$

The definition (equation A 13.20) of $\sigma_{\alpha\beta}$ has been used to obtain the final expressions and the three constants are quoted in Voigts' notation.[10]

It should be noted that in calculating the Fourier transformation of a stress correlation function the asymptotic part is first subtracted off. This avoids a $\delta(Q)$ term, and also confirms that the elastic moduli are related to the fluctuating part of the stress correlation, a fact which is indicated by the *prime* on the stress tensor components.

Symbols for Chapter 13

a	Coefficient in equation (13.19)	$P_{\alpha\beta}$	Force per unit area exerted in the α direction across a surface perpendicular to the β direction (a prime denotes fluctuating part)
A_i	Operator defined by equation (13.31b)		
B_0	Bulk modulus defined by equation (13.30)		
B_1	"Relaxation" bulk modulus	$s_i(t)$	Deviation of Δ^V/V from its equilibrium value
E	Electric field of electromagnetic wave	u	Strain tensor
E	Internal energy of liquid	$\alpha, \beta, \gamma, \delta$	Cartesian co-ordinate labels
j	Atomic current density	η'	Dilational viscosity
l	A displacement	Λ	Rate of strain tensor
		τ_m	Maxwell relaxation time

CHAPTER 14

Co-operative Modes of Motion at High Frequencies

The discussion of Chapter 13 was centred on the $Q \to 0$ region of (ω, Q) space where the liquid may be treated as a visco-elastic continuum. In this Chapter, larger values of Q and ω are considered that are relevant to the high-frequency modes of motion in a liquid. At first, this treatment will follow the method used for the discussion of single-particle modes, in that the viscosity coefficient will be expressed as the time integral of a correlation function and the high-frequency behaviour of that function outlined. It will be shown that it (like the velocity correlation function) is a $Q \to 0$ limit, but unlike the single-particle case it is not possible to relate the $Q \neq 0$ behaviour to the $Q \to 0$ behaviour in an approximate way. A comparison with the case of a solid is instructive here, since it is easy to show that for a crystal the low wave-number phonons are not simply related to the phonons of high wave number. In contrast, the frequency spectrum of modes (which gives the average motion of a single atom) contains information about all the phonons even though it is observed as a $Q \to 0$ limit.

Thus for the co-operative modes the $Q \neq 0$ region will be given a separate treatment. Unfortunately a complete theory is not available, and therefore this discussion will be incomplete. In a complete theory of liquids, the single-particle and co-operative modes of motion would be combined into one larger treatment. A number of unsuccessful attempts to do this have been made, and an example of this kind will be discussed at the conclusion of this Chapter.

14.1. Stress Correlation Functions

It may be shown rigorously,[5] that the viscosity is a time integral of the fluctuating part of the stress correlation function, and a plausibility argument for this result will be given now.

The viscosity coefficient is usually referred to as the "diffusion coefficient" for momentum, a point that may be illustrated by combining the momentum conservation condition (equation 13.13 or 13.33) with the definition of viscosity (equation A 13.2). It is imagined that forces are applied to the liquid that set it flowing. Then these forces are switched off, and the equation of motion is reconsidered. As an example, consider the shear viscosity alone (i.e. put $\alpha \neq \beta$ in equation A 13.2), and insert the definition of viscosity into equation (13.13), which then reduces to—

$$\frac{\partial v_\alpha}{\partial t} = \frac{\eta}{M\rho} \nabla^2 v_\alpha \tag{14.1a}$$

This is a diffusion equation for the momentum including a diffusion constant equal to $\eta/M\rho$. Thus, following the discussion of diffusion in Chapter 10, there will be a relationship for η of the form—

$$\eta/M\rho = \overline{l^2}/6\tau_0 \tag{14.1b}$$

The distance l is measured, for this case, in terms of the velocity gradient—

$$l^2 \equiv (\Delta v_\alpha)^2 / \nabla^2 v_\alpha$$

or in terms of the momentum space discussion used in Chapter 13—

$$\overline{l^2} \equiv \left\{ \frac{\langle(\Delta v_\alpha)_{-Q}(\Delta v_\alpha)_{+Q}\rangle}{\langle(\nabla v_\alpha)_{-Q}(\nabla v_\alpha)_{+Q}\rangle} \right\}_{Q\to 0} \tag{14.2a}$$

on using the generalization of equation (11.7), and transforming the denominator—

$$= \left\{ \frac{6\tau_0 \int_0^\infty \left\langle \left(\frac{\partial v_\alpha(0)}{\partial t}\right)_{-Q} \cdot \left(\frac{\partial v_\alpha(\tau)}{\partial t}\right)_{+Q} \right\rangle d\tau}{\langle Q^2(v_\alpha)_{-Q}(v_\alpha)_{+Q}\rangle} \right\}_{Q\to 0} \tag{14.2b}$$

on using equation (13.34) and equation (A 13.25)—

$$= \frac{6\tau_0}{NMkT} \int_0^\infty \langle \{\sigma'_{\alpha\beta}(0)\sigma'_{\alpha\beta}(\tau)\}_{Q\to 0}\rangle d\tau \tag{14.2c}$$

At this point the $Q \to 0$ notation will be dropped, as this limit will apply to all correlation functions in this Chapter. Also, the convention of Appendix A4 (Chapter 13) will be used in which the Q direction is taken as the z direction (note the convention used in Chapter 11, which is conventional in the discussion of $G_s(\mathbf{r}, \tau)$, was that the Q direction was in the x direction). In this case, equations (14.2c) and (14.1b) may be combined to give—

$$\eta = \frac{1}{VkT} \int_0^\infty \langle \sigma'_{zx}(0)\sigma'_{zx}(\tau)\rangle d\tau \tag{14.3a}$$

A similar discussion for the quantity $(\tfrac{4}{3}\eta + \zeta)$ gives the result—

$$\tfrac{4}{3}\eta + \zeta = \frac{1}{VkT} \int_0^\infty \langle \sigma'_{zz}(0)\sigma'_{zz}(\tau) \rangle \, d\tau \qquad (14.3b)$$

As in the discussion of Chapter 11, spectral density functions are defined as the Fourier transform of the stress correlation function, or—

$$\left.\begin{aligned} z_\eta(\tau) &= \frac{1}{V} \langle \sigma'_{zx}(0)\sigma'_{zx}(\tau) \rangle \\ \tilde{z}_\eta(\omega) &= \frac{1}{2\pi} \int_{-\infty}^{+\infty} e^{-i\omega\tau} z_\eta(\tau) \, d\tau \end{aligned}\right\} \qquad (14.3c)$$

and in this case—

$$\tilde{z}_\eta(0) = \frac{\eta kT}{\pi}$$

Also—

$$\int_{-\infty}^{+\infty} \tilde{z}_\eta(\omega) \, d\omega = kTG$$

from equation (A 13.27c); and the remark following that equation should be noted, namely that only the fluctuating part of the stress correlation function is used in these expressions (indicated by a *prime* on σ). Time-independent parts of the stress correlation (proportional to $1/\chi_s$) are excluded in order to exclude a $\delta(\omega)$ term from the spectral density of the (zz) correlation function. The (zx) correlation falls to zero at long times as shown in the next Section.

14.2. Stress Correlation in the Visco-elastic Theory

As an example of a stress correlation function consider the shear viscosity in the visco-elastic theory. If r_z is the elastic displacement along the z axis then—

$$\frac{\partial r_z}{\partial x} = \frac{\sigma_{zx}}{G} \qquad (14.4a)$$

and by differentiating with respect to t—

$$\frac{\partial v'_z}{\partial x} = \frac{1}{G} \cdot \frac{\partial \sigma_{zx}}{\partial t} \qquad (14.4b)$$

where $v'_z = \partial r_z/\partial t$. Let viscous flow take place with a velocity v''_z, then from the definition of a viscosity coefficient it is seen that—

$$\frac{\partial v_z''}{\partial x} = \frac{\sigma_{zx}}{\eta} \qquad (14.4c)$$

The addition of equations (14.3a) and (14.3b) gives the visco-elastic equation for the total velocity $v_z = v_z' + v_z''$—

$$\frac{\partial v_z}{\partial x} = \frac{1}{G}\frac{d\sigma_{zx}}{dt} + \frac{\sigma_{zx}}{\eta} \qquad (14.5)$$

(note: this equation is equivalent to equation 13.16, if v_z and σ_{zx} have an oscillatory form). In order to calculate the stress correlation function, it is necessary to stop the motion suddenly (i.e. put $v_z = 0$) and calculate how σ_{zx} relaxes. Thus from equation (14.5) with $v_z = 0$—

$$\sigma_{zx}(\tau) = \sigma_{zx}(0)\, e^{-\tau/\tau_m}$$

Now if this equation is multiplied by $\sigma_{zx}(0)$ and thermally averaged, it is found that—

$$\frac{1}{V}\langle \sigma_{zx}(0)\,.\,\sigma_{zx}(\tau)\rangle = \frac{1}{V}\langle \sigma_{zx}^2(0)\rangle\, e^{-\tau/\tau_m}$$

on using equation (13.27c)—

$$= kTG\, e^{-\tau/\tau_m} \qquad (14.6a)$$

Fourier transformation of equation (14.6a) gives a spectral density of Lorentzian form—

$$\tilde{z}_\eta(\omega) = \frac{kTG}{\pi}\frac{\tau_m}{1+\omega^2\tau_m^2} \qquad (14.6b)$$

Figures 14.1(a) and (b) show the form of equation (14.6) and the corresponding spectral densities. This expression is very similar

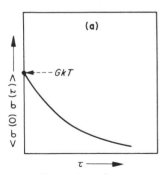
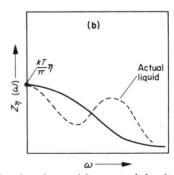

FIG. 14.1. Illustration of a stress correlation function and its spectral density. The form of equation (14.6a) is shown in (a) and (14.6b) in (b).

to the equation (11.29b) for the velocity correlation function and suffers from the same defects. It may be generalized to the form of equation (11.29c), but this does not remove the discrepancy shown in Fig. 14.1(b).

An equation analogous to equation (14.6) can be derived for the bulk viscosity also. Equations such as these can be regarded as having the correct long time (or low frequency) behaviour but being inadequate at high frequencies as shown in Fig. 14.1(b).

14.3. Comments on Viscosity Formulae

The spectral density of the stress correlation function provides a useful concept for the discussion of equations (12.21) and other viscosity formula. For example, equation (12.21a) may be rewritten in terms of the integrals I_1 and I_2 (equation 13.38)—

$$\eta = \frac{\rho kT}{15\gamma}(I_2 + 4I_1) = \frac{c_{44}}{\gamma} - \frac{\rho kT}{15\gamma} \simeq G/\gamma \qquad (14.7a)$$

The $\rho kT/15\gamma$ term in this expression is the kinetic term, which was ignored in the derivation of η in equation (12.20). Noting that the area of the spectral density function is kTG it becomes clear that equation (14.7a) reduces to—

$$\eta \equiv \frac{\text{Area}}{\text{Width}} \text{ of spectral density} \qquad (14.7b)$$

which is consistent with the fact that η is the $\omega = 0$ intercept of the spectral density (Fig. 14.1b) and an assumption that $\tilde{z}_\eta(\omega)$ has a simple form such as equation (14.6b). However for a realistic form for $\tilde{z}_\eta(\omega)$ this expression suffers from the same defects as the expression for γ discussed in Chapter 12 (equation 12.14). It is probable, however, that the function $\tilde{z}_\eta(\omega)$ is nearer to equation (14.6b) than $\tilde{z}(\omega)$ is to equation (11.29b), and hence the errors in this type of treatment are less important for the viscosity coefficient than the diffusion coefficient.

Green[53] uses another method of calculating the viscosity, based on an evaluation of the changes in $g(r)$ which occur when a liquid is flowing. His result is—

$$\left.\begin{aligned} \eta &= \frac{\rho^2}{30\gamma_1} \int_V rg(r)\frac{du(r)}{dr}\,dr = \frac{\rho kT}{15\gamma_1} I_1 \\ \eta' &= \zeta - \tfrac{2}{3}\eta = \frac{\rho^2}{2\gamma_1'} \int_V rg(r)\frac{du(r)}{dr}\,dr = \frac{\rho kT}{\gamma_1'} I_1 \end{aligned}\right\} \qquad (14.8)$$

where γ_1 and γ_1' are frequencies of the order of the frequency of oscillation in the liquid. These expressions have the same form as equation (14.7), but include fewer terms and the frequencies γ_1 and γ_1' are not well defined. Thus these expressions can be interpreted in the same way as equation (14.7b) (note however that if $\gamma \sim \gamma_1$, the ratio of η from equation (14.7a) to η from equation (14.8) is $I_2/I_1 \sim 100$). Equations such as (14.7) and (14.8) can be made more realistic by using the idea of an "effective modulus" for viscosity and interpreting the frequency γ as the (life-time)$^{-1}$ for a viscous movement. In this case the viscosity cannot be calculated without an independent estimate of these parameters. This interpretation is analogous to the effective mass and friction constant relationship in diffusion theory (equation 10.20).

14.4. Measurement of the (z, z) Stress Correlation Function

The stress correlation can be measured (in principle) by taking a limit to the scattering law $S(Q, \omega)$. This can be shown conveniently from the definition of the intermediate scattering function $I(Q, \omega)$—equation (11.23)—i.e.—

$$I(Q, \tau) = \frac{\langle \rho(-Q, 0)\rho(Q, \tau) \rangle}{N} \tag{14.9}$$

If this equation is differentiated twice with respect to τ, and the stationarity condition is used to simplify the result, it may be shown that—

$$\frac{\partial^2 I(Q, \tau)}{\partial \tau^2} = -\frac{1}{N} \langle (\mathbf{j}_{-Q}(0) . \mathbf{Q})(\mathbf{j}_Q(\tau) . \mathbf{Q}) \rangle \tag{14.10}$$

where j_Q is defined at equation (13.32). Then equation (14.10) is differentiated twice with respect to τ (and again the stationarity condition is used) so giving the final result—

$$\frac{\partial^4 I(Q, \tau)}{\partial \tau^4} = \frac{1}{\rho^2} \langle (\mathbf{Q} . \boldsymbol{\sigma}_{-Q}(0) . \mathbf{Q})(\mathbf{Q} . \boldsymbol{\sigma}_Q(\tau) . \mathbf{Q}) \rangle$$

$$= \frac{Q^4}{\rho^2} \langle (\sigma_{zz}(0))_{-Q}(\sigma_{zz}(\tau))_Q \rangle \tag{14.11}$$

where the z axis is along Q and $(\sigma_{\alpha\beta})_Q$ is defined by equation (A 13.16). Finally, equation (14.11) is divided by Q^4 and the limit $Q \to 0$ is taken, or—

$$\left[\frac{1}{Q^4} \frac{\partial^4 I(Q, \tau)}{\partial \tau^4} \right]_{Q \to 0} = \frac{1}{\rho^2} \langle \sigma_{zz}'(0)\sigma_{zz}'(\tau) \rangle \tag{14.12a}$$

and on Fourier-transforming this equation, it is found that—

$$\omega^4\left[\frac{S(Q,\omega)}{Q^4}\right]_{Q\to0} = \frac{1}{2\pi}\int_{-\infty}^{\infty} e^{-i\omega\tau}\left[\frac{\langle\sigma'_{zz}(0)\sigma'_{zz}(\tau)\rangle}{\rho^2}\right]d\tau = \tilde{z}_V(\omega) \quad (14.12b)$$

The difficulty in using this equation to obtain measured values of $\tilde{z}_V(\omega)$ lies in the problem of extrapolating the experimental data to low values of Q. This problem is illustrated in Fig. 14.2, which compares the type of extrapolation needed for equation (14.12b) with that needed for equation (11.20). Because of these difficulties, adequate data on $\tilde{z}_V(\omega)$ have not been obtained yet.

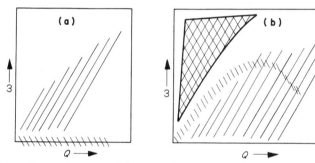

FIG. 14.2. Comparison of problems in using equations (11.20) and (14.12b) to obtain spectral densities. Diagram (a) shows the position of the ridge in $S_s(Q,\omega)$—\\\\\\—compared to the region covered by neutron-scattering experiments—////. Diagram (b) shows the position of the ridge in $S(Q,\omega)$—\\\\\—compared to the region covered by neutron-scattering experiments—////. In this case, extrapolation through the ridge is required, and to avoid this, measurements are needed in the cross-hatched —XXXX—region. (From Egelstaff,[42] Fig. 5.)

It is clear from equation (14.12b) that the fourth ω moment of $S(Q,\omega)$ should be related to the elastic constants. Equation (11.25) has to be evaluated for all (ij). The use of equation (11.26) simplifies equation (11.25), giving—

$$\int_{-\infty}^{\infty} \omega^4 S(Q,\omega)\,d\omega = 3Q^4\left(\frac{kT}{M}\right)^2 + \frac{Q^2kT}{M^2N}\sum_{ij}\left\langle\frac{\partial^2 U\{N\}}{\partial z_i\partial z_j}e^{iQz_{ij}}\right\rangle \quad (14.13a)$$

Finally through equation (2.17), the fourth moment becomes—

$$\frac{1}{Q^4}\int_{-\infty}^{+\infty} \omega^4 S(Q,\omega)\,d\omega = 3\left(\frac{kT}{M}\right)^2 + \frac{\rho kT}{M^2}\int g(r)\left[\frac{1-\cos Qz}{Q^2}\right]\frac{\partial^2 u(r)}{\partial z^2}\,d\mathbf{r} \quad (14.13b)$$

which in the $Q\to0$ limit is equal to $kTc_{11}/M^2\rho$ from equation (A 13.27a). This confirms that the area of $\tilde{z}_V(\omega)$ is related to an elastic constant.

14.5. Comments on the Thermal Conductivity

The thermal conductivity is defined as the ratio of the flux of thermal energy to the temperature gradient in a volume of liquid over which a uniform temperature gradient has been established. In this Section, the correlation function definition of thermal conductivity will be given, and its significance as the $Q \to 0$ limit of a part of $S(Q, \omega)$ will be considered.

It may be shown (e.g. Rice and Gray,[5] p. 546) that the thermal conductivity may be expressed as the time integral of a heat current correlation function, i.e.—

$$\lambda = \frac{1}{VkT^2} \int_0^\infty \langle J_x(0) . J_x(\tau) \rangle \, d\tau \qquad (14.14a)$$

where—

$$J_x(\tau) = \frac{d}{d\tau} \sum_i r_{ix}(\tau) \tilde{E}_i$$

and $\tilde{E}_i$ is the fluctuation of the energy of the ith atom from its average value. This correlation function may be measured in the continuum region if the total scattering law may be broken down into entropy and pressure fluctuation parts. In principle, this is possible by use of the fact that (Chapter 8) the scattering of neutrons is related to density fluctuations, whereas the scattering of light is related to the dielectric-constant fluctuations. These functions contain the two components in different ratios so that the observed intensity may be separated, in principle, into the entropy and pressure-fluctuation parts. This method has never been used because existing techniques are not sufficiently sensitive. However in some cases a separation is possible on the basis of frequency range alone (as shown in Fig. 14.5a), and sometimes the entropy fluctuations are the dominant contribution to $S(Q, \omega)$, so that the pressure-fluctuation term may be neglected (e.g. at the critical point as in Fig. 16.5a).

If the part of $S(Q, \omega)$ due to the entropy fluctuations—denoted by $S_T(Q, \omega)$—is considered separately, it may be shown to have the limiting form—

$$S_T(Q, \omega) \simeq \rho kT\chi_T \frac{C_P - C_V}{C_P} . \frac{1}{\pi} \frac{D_T'Q^2}{(D_T'Q^2)^2 + \omega^2} \quad \text{for } Q \sim 0 \qquad (14.14b)$$

and—

$$D_T' = \lambda/C_P\rho$$

The first factor follows from equation (13.46) and the expression involving D'_T follows from equation (13.11), and the thermal conductivity is given as the limit—

$$\frac{C_P^2}{(C_P - C_V)\chi_T} \cdot \left\{\omega^2 \left[\frac{S_T(Q, \omega)}{Q^2}\right]_{Q \to 0}\right\}_{\omega \to 0} \to \frac{\lambda k T}{\pi} \qquad (14.14c)$$

This expression is analogous to the expression $\tilde{z}(0)$ for the diffusion coefficient and $\tilde{z}_\eta(0)$ for the viscosity coefficient. It consists of two factors, the first normalizes the amplitudes of the density and entropy fluctuations, and the second is analogous to the earlier expressions.

The heat current correlation function is expected to have the same general form as the stress correlation shown in Fig. 14.1b. Owing to the experimental difficulties there are almost no data available concerning its behaviour.

14.6. $\mathscr{I}\{G(\mathbf{r}, \tau)\}$ for a Classical Liquid

It has been pointed out (equation 8.30) that the imaginary part of $G(\mathbf{r}, \tau)$ divided by $\hbar$ has a physical meaning in a classical liquid. As an illustration of this, $S(Q, \omega)$ may be written in terms of $\mathscr{I}\{G(\mathbf{r}, \tau)\}$ by using equation (8.27). It is convenient to write this result in terms of a new quantity $A(Q, \omega)$, defined as—

$$A(Q, \omega) = \frac{2kT}{\hbar} \int_0^\infty \mathscr{I}\{I(\mathbf{Q}, \tau)\} \, e^{-i\omega\tau} \, d\tau \qquad (14.15a)$$

where $I(Q, \tau)$ is the intermediate scattering function (equation 9.2). It should be noted that the integral in this equation extends from zero to infinity, in contrast to the integral from minus to plus infinity in equation (8.10), for example. Thus—

$$S(Q, \omega) = \frac{1}{2\pi} \int_{-\infty}^{+\infty} \{\mathscr{R}[I(Q, \tau)] + i\mathscr{I}[I(Q, \tau)]\} \, e^{-i\omega\tau} \, d\tau$$

on using equations (8.26) and (8.27)—

$$= \left[\coth\frac{\hbar\omega}{2kT} + 1\right] \frac{i}{2\pi} \int_{-\infty}^{+\infty} \mathscr{I}\{I(Q, \tau)\} \, e^{-i\omega\tau} \, d\tau$$

$$= \frac{\hbar}{2\pi kT} \frac{e^{\hbar\omega/2kT}}{\sinh(\hbar\omega/2kT)} \mathscr{I}\{A(Q, \omega)\} \qquad (14.16a)$$

$$\simeq \frac{1}{\pi} \frac{\mathscr{I}\{A(Q, \omega)\}}{\omega} \qquad (14.16b)$$

in the classical limit.

The physical significance of $A(Q, \omega)$ may be gathered from van Hove's[54] description of Imag $G(\mathbf{r}, \tau)$, in the form employed by Sjolander,[55] i.e.—

$$\frac{\rho - \langle \rho(\mathbf{R}, t) \rangle_F}{\rho} = \int_{-\infty}^{t} d\tau \int d\mathbf{R} \left\{ \frac{2}{\hbar} \mathscr{I}[G(R - \mathbf{r}, t - \tau)] V(\mathbf{R}, t) \right\} \quad (14.17)$$

where $V(\mathbf{R}, t)$ is an externally applied potential, and the subscript F indicates that the average is taken in the presence of the applied field. Thus Imag G is related to the density fluctuations caused by the applied potential. Now if V is an oscillatory potential of the form—

$$V(\mathbf{R}, t) = kT \exp - i\{\mathbf{Q} . (\mathbf{R} - \mathbf{r}) - \omega(t - \tau)\} \quad (14.18)$$

and equation (14.18) is substituted into equation (14.17), it is found that the $\mathbf{R} \to 0$, $t \to 0$ limit gives from (14.15a)—

$$\left[\frac{\langle \rho(0, 0) \rangle_F - \rho}{\rho} \right]_{Q, \omega} = A(Q, \omega) \quad (14.19)$$

Thus $A(Q, \omega)$ is the amplitude of a Fourier mode applied at the origin at $t = 0$.

Another description of $\mathscr{I}G(\mathbf{r}, \tau)$ was given by Egelstaff,[56] who showed from the continuity equation (13.2) and equation (8.30) that, in a classical liquid—

$$\frac{2kT}{\hbar} \mathscr{I}\{G(\mathbf{r}, \tau)\} = \frac{1}{\rho} \nabla . \langle \rho(0, 0) . \mathbf{j}(\mathbf{r}, \tau) \rangle \quad (14.20a)$$

where $\mathbf{j}$ is given by equation (13.42a). Thus on Fourier-transforming—

$$i\mathbf{Q} . \mathbf{P}(Q, \omega) = A(Q, \omega) \quad (14.20b)$$

where—

$$\mathbf{P}(Q, \omega) = \frac{1}{2\pi} \int e^{i(\mathbf{Q}.\mathbf{r} - \omega\tau)} \frac{\langle \rho(0, 0)\mathbf{j}(\mathbf{r}, \tau) \rangle}{\rho} d\mathbf{r} \, d\tau \quad (14.20c)$$

In this case $A(Q, \omega)$ is the projection onto $\mathbf{Q}$ of the correlation between the current at $(\mathbf{r}, \tau)$ and the amplitude of a density fluctuation at $(0, 0)$. In this sense, $A(Q, \omega)$ is a function describing the transport of matter. The equivalence of equations (14.20b) and (14.19) is a statement of the fluctuation–dissipation theorem in Fourier space.

14.7. Frequency Wave-number Relation for High-frequency Modes

Because of the physical meaning of $A(Q, \omega)$ several authors[56] suggested that it should satisfy the usual equation of motion for longitudinal waves. In the limit of low frequency ($Q \rightarrow 0$), this would be exactly true. This equation is written—

$$\left\{\frac{\partial^2}{\partial t^2} - \frac{V(Q, \omega)}{\rho}\frac{\partial}{\partial t}\nabla^2 - \frac{kT}{MC}\nabla^2\right\}\{A(Q, \omega)\exp \text{i}(\mathbf{Q}.\mathbf{r} - \omega\tau)\}$$

$$= B \exp \text{i}(\mathbf{Q}.\mathbf{r} - \omega\tau) \tag{14.21}$$

where B and C are constants to be determined later, and $V(Q, \omega)$ is a general transport coefficient.[57] The term on the right-hand side is a driving force; in the present context it represents the random forces in the liquid which drive a thermally excited sound wave of wave number Q and frequency ω. Equation (14.21) is readily solved for $A(Q, \omega)$ giving—

$$A(Q, \omega) = \frac{MB/kT}{Q^2/C - [\text{i}\omega MQ^2 V(Q, \omega)/kT] - M\omega^2/kT} \tag{14.22}$$

Since $A(Q, \omega)$—equation (14.15a)—is a one-sided Fourier transform, its real and imaginary parts are related by a Kramers–Kronig relation, which in this case is written—

$$\mathscr{R}\{A(Q, \omega)\} - \mathscr{R}\{A(Q, 0)\} = P\!\int_{-\infty}^{+\infty} \frac{\omega\,\text{d}\omega'}{\pi}\frac{\mathscr{I}\{A(Q, \omega')\}}{(\omega' - \omega)\omega'} \tag{14.23}$$

where P denotes principal part. The right-hand side of this equation may be expanded as power series in $1/\omega$ (i.e. an expansion for $\omega \rightarrow \infty$), giving—

$$\mathscr{R}\{A(Q, \omega)\} - A(Q, 0) = -\frac{1}{\pi}P\!\int_{-\infty}^{\infty}\frac{\mathscr{I}\{A(Q, \omega')\}}{\omega'}\sum_{n=0}^{\infty}\left(\frac{\omega'}{\omega}\right)^n\text{d}\omega' \tag{14.24}$$

From equation (14.16b), the right-hand side of this equation is given immediately in terms of the ω moments of $S(Q, \omega)$. In this way the coefficients B and C are found to be—

$$\left.\begin{array}{l}B = \dfrac{kT}{M}Q^2 \\[2mm] C = S(Q) = A(Q, 0)\end{array}\right\} \tag{14.25}$$

[Note: the latter equation states that $g(r)$ is the time integral of the imaginary part of $G(r, \tau)$.]

Thus the classical scattering law is found from equation (14.16b) to be—

$$S(Q, \omega) = \frac{1}{\pi}\frac{Q^4\mathscr{R}\{V(Q, \omega)\}}{|[Q^2 kT/MS(Q)] - \text{i}\omega Q^2 V(Q, \omega) - \omega^2|^2} \tag{14.26}$$

and the poles of this equation give the frequency wave-number relationship, or—

$$\omega^2 = \frac{Q^2 kT}{MS(Q)} - i\omega Q^2 V(Q, \omega) \tag{14.27}$$

As expected, the latter equation has the same form as equation (13.27). Consequently the $Q \to 0$ expression for $V(Q, \omega)$ is obtained directly from a comparison of these two equations and Table (13.1). Equation (14.12b) shows that $V(0, \omega)$ is equal to $\tilde{z}_V(\omega)$, i.e.—

$$\mathcal{R}\{V(0, \omega)\} = \frac{1}{\pi \rho^2} \int_0^\infty \cos \omega\tau \langle \sigma'_{zz}(0)\sigma'_{zz}(\tau) \rangle \, d\tau \tag{14.28}$$

At non-zero values of Q, the form of $V(Q, \omega)$ is unknown theoretically, but can be determined by a comparison of experimental values of $S(Q, \omega)$ with equation (14.26). A simple way of doing this is to plot a graph of the (ω–Q) values of the observed peaks in $S(Q, \omega)$ and compare these data with equation (14.27). Figure (14.3) shows the data[56] on lead compared to a simplified form of equation (14.27), i.e.—

$$\omega_m^2 = N^2 \frac{Q^2 kT}{MS(Q)} \tag{14.29}$$

where N is a constant. It can be seen from the figure that, at the highest

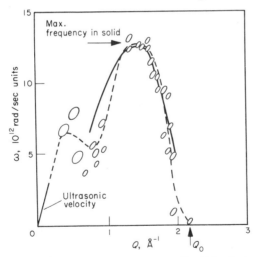

FIG. 14.3. Graph of observed peaks in $S(Q, \omega)$ for liquid lead; the error is indicated by the size of the ellipse marking each point. The lower full line indicates the extrapolation of the ultrasonic region, the dotted line indicates a possible form for the curve, and the upper full line is equation (14.29) for $N = 2.3$. (From Egelstaff,[56] Fig. 6.)

values of ω, a reasonable fit is obtained with N equal to 2·3, approximately. However at low values of ω the frequency dependence of $V(Q, \omega)$ is sufficiently important to render the approximation (14.29) invalid. It is interesting to compare equation (14.29) with equation (13.27) written in the limit of $\omega \to \infty$ as—

$$\omega^2 \to \frac{Q^2 kT}{MS(O)} \left[\chi_T(B + \tfrac{4}{3}G) \right] \qquad (14.30)$$

In a typical case, the factor in square brackets amounts to $\sim 1·2$, so that the variation with Q has an important effect on this term. To illustrate this point Fig. 14.4 gives the ratio of observed (ω_{obs}) to theoretical frequencies—

$$R(Q) = \frac{(\omega_{obs})^2 MS(Q)}{Q^2 kT} = N^2 \qquad (14.31)$$

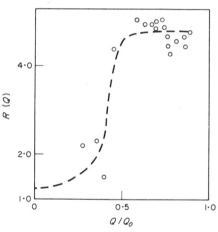

FIG. 14.4. The ratio $R(Q)$—equation (14.31)—for liquid lead; at $Q \to 0$, the value may be calculated from the results in Table 13.1, and the pronounced variation with Q (shown by the dashed line which is drawn by eye through the experimental points) reflects the behaviour of $V(Q, \omega)$—equation (14.27). (From Egelstaff,[42] Fig. 8.)

as a function of Q. If the real part* of $V(Q, \omega)$ were much smaller than the real part of equation (14.27) (so that $S(Q, \omega)$ consisted of three sharp lines), then this ratio could be calculated from the fourth moment of $S(Q, \omega)$ as—

$$R(Q) = S(Q) \left[3 + \frac{\rho}{kT} \int_V g(r) \left[\frac{1 - \cos Qz}{Q^2} \right] \frac{\partial^2 u(r)}{\partial z^2} \, d\mathbf{r} \right] \qquad (14.32)$$

Values of this expression have not so far been computed.

* The real and imaginary parts are related by the Kramers–Kronig relation.

The similarity in the maximum frequency observed for solid and liquid metals (e.g. Fig. 14.3) and for rare-gas liquids[56] indicates that neighbouring atoms are found at a similar spacing in both states. It should be noted that this fact cannot be deduced from $g(r)$ alone, because $g(r)$ does not refer to the "equilibrium" positions about which the atoms are vibrating.

14.8. Lifetime of Co-operative Modes

The behaviour of the co-operative modes as a function of Q depends upon the size of $\mathscr{R}\{V(Q, \omega)\}$. As an example, the polycrystalline solid may be considered, since here the modes have an infinite lifetime and $\mathscr{R}\{V(Q, \omega)\}$ is zero. At low Q, $S(Q, \omega)$ exhibits sharp peaks at values of $\omega = cQ$ where c is the velocity of sound. With increasing Q, the dispersion of longitudinal modes is observed, and after passing from the first Brillouin zone into the second, a broad spectrum develops. This spectrum is displaced from the origin and may contain several peaks: it arises because of "umklapp" processes in which momentum is shared between a high-frequency phonon and an $\omega = 0$ or translational mode. These processes allow both longitudinal and transverse phonons to contribute to $S(Q, \omega)$. In this region (and at higher Q also) multi-phonon processes, in which two or more high-frequency phonons contribute to $S(Q, \omega)$, become an important factor in determining the width of the spectrum. At very high values of Q, the scattering law reduces to the perfect-gas form (Chapter 9) and becomes a single-peaked curve centred on $\omega = 0$. Density fluctuations at these high values of Q are due to single-particle modes rather than co-operative modes.

For a liquid, the behaviour of $S(Q, \omega)$ as a function of Q is rather similar, and is illustrated in Fig. 14.5. At low Q (Fig. 14.5a), the sound-wave modes are clearly defined, but in addition there is a central peak corresponding to the entropy fluctuations (equation 13.4b). These three peaks are known as the Brillouin doublet and Rayleigh line, respectively, in optical work. As Q is increased, the width of the lines increases due to the increasing size of $\mathscr{R}V(Q, \omega)$. The lifetime of the co-operative mode (τ_L) is equal to—

$$\tau_L \sim \frac{2\hbar}{Q^2\mathscr{R}\{V(Q, \omega)\}} \tag{14.33}$$

from equation (14.27) for small values of $V(Q, \omega)$, and hence decreases with increasing values of $\mathscr{R}V(Q, \omega)$. The experimental data[56] shows that the lifetime is great enough for observable maxima to be found

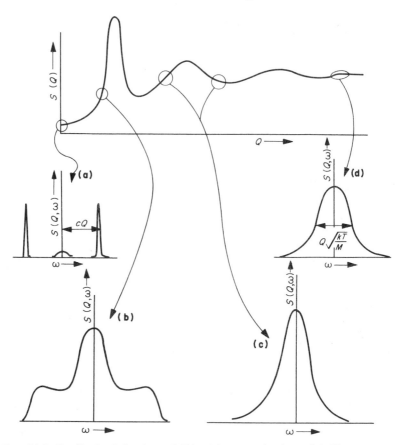

Fig. 14.5. Qualitative behaviour of $S(Q, \omega)$ for several values of Q. The upper curve is the function $S(Q)$, and the remaining curves show the spectral shape at fixed values of Q marked by circles on $S(Q)$. At low and high values of Q, the width can be calculated from simple considerations.

in $S(Q, \omega)$ for $Q \lesssim Q_0$ (Fig. 14.5b). These maxima have a width, at the highest observed value of ω, that is not very different from the value for the crystalline solid. For example, in lead for a mode of frequency 12.5×10^{12} cycles/sec, the lifetime is $\sim 0.3 \times 10^{-12}$ sec in the liquid and $\sim 0.5 \times 10^{-12}$ sec in the solid at the melting point. This similarity arises because the highest frequencies correspond to adjacent atoms moving in opposition—in the solid their Q values correspond to the boundary between the first and second Brillouin zones. It is found experimentally that for crystals near the melting point, the mean free path (i.e. lifetime multiplied by propagation velocity) is only a few atomic spacings. Since this is roughly the range of the order in a liquid [deduced from the oscillations in $g(r)$] a mode of this type will be damped before the difference in order between the solid and liquid becomes significant.

At $Q \sim Q_0$, the function $S(Q, \omega)$ consists of a single, relatively sharp, peak centred on $\omega = 0$ (i.e. a non-propagating mode). The width of this peak is roughly equal to the lifetime for a group of atoms to act together, because adjacent atoms are moving in phase in this case. Experimentally, the life-times are found to be $\sim 10^{-12}$ sec, which is of the same order as the time for a diffusive step (Table 10.1). The full significance of this result is not yet clear.

At higher values of Q, the maxima in $S(Q, \omega)$ smear out, and eventually the scattering law becomes a single-peaked function centred on $\omega = 0$ (Fig. 14.5c). However the width, $\Delta\omega$, may not be equal to the width for a perfect gas, because a single-peaked function can be related to a large value of $\mathcal{R}V(Q, \omega)$—that is modes with very short life-times. It is found that (for metals) high values of $Q \sim 10\,\text{Å}^{-1}$, are required before the gas limit is reached (Fig. 14.5d).

14.9. Combined Single-particle and Co-operative Mode Schemes

One of the basic questions of liquid-state theory is, how may the co-operative and single-particle modes be included in one treatment? The above discussion of $S(Q, \omega)$ is a qualitative answer to this question; a detailed answer cannot be given at the present time. Nevertheless, there have been several attempts at an answer, one of the simplest being due to Vineyard.[36] He suggested that (see Fig. 8.1) an atom placed at 0 might move over the length $\mathbf{r}'$ in time τ, independently of the atom at $\mathbf{r}' - \mathbf{r}$. In this case, the total correlation function is simply—

$$G(\mathbf{r}, \tau) \simeq G_S(\mathbf{r}, \tau) + \rho \int_V g(|\mathbf{r}' - \mathbf{r}|) G_S(\mathbf{r}', \tau)\, d\mathbf{r}' \qquad (14.34)$$

Fourier transformation of this equation gives the result—

$$S(Q, \omega) \simeq S(Q) S_S(Q, \omega) \qquad (14.35)$$

It is easy to see that although this equation satisfies the sum rule (equation 8.22), it fails to satisfy equation 9.11. Also, because of the difference in shape (at low Q) between $S_S(Q, \omega)$—Fig. 10.4(a)—and the complete S function (Fig. 14.5a), equation (14.35) does not even reproduce the correct qualitative features of $S(Q, \omega)$. The major defect in this model is that it allows a finite chance for two atoms to be at $\mathbf{r}'$ at the same time. A variety of methods have been suggested to improve it, some mathematical and some physical in character, but they have not led to a satisfactory solution.

Another method is to try to relate the functions $V(Q, \omega)$, and $\tilde{z}(\omega)$ —equations (14.28) and (11.10), respectively. This is rather similar to the problem of relating the diffusion and viscosity coefficients discussed in Chapter 12. It has the advantage that a physical model can be introduced more easily than in the former method. For example, in the case of a hard-sphere gas (equation 12.20), the relationship is—

$$\frac{\mathcal{R}\{V(0, \omega)\}}{3MG} = \frac{1}{\pi}\frac{kT}{M}\frac{\tau_m}{1+\omega^2\tau_m^2} = \tilde{z}(\omega) \qquad (14.36)$$

since (equation A 13.27) $3B/5 = G = \rho kT$ in a gas, and thus $MD/kT = \eta/G$. At non-zero values of Q, the relation is more complex. Significant progress in extending such relationships to the liquid state has not been made so far.

The discussion of this Chapter has covered the properties of high-frequency modes in broad outline. Detailed behaviour of these properties is difficult to predict, because the theory has not yet reached the point of yielding rigorous expressions for them in terms of numerically accessible quantities.

Symbols for Chapter 14

$R(Q)$ Ratio of calculated to observed frequencies (equation 14.31)
τ_L Life-time of co-operative mode
ω_m Life-time of co-operative mode

CHAPTER 15

The Liquid–Gas Critical Point

The discussion of the liquid state given in the previous fourteen Chapters has concerned primarily the region marked by shading in Fig. 1.1. To conclude this introduction, a brief outline will be given of the extreme limits of the liquid state, the critical point on the one hand and the superfluid region on the other. The properties of liquids in these regions should be compared to those of "normal" liquids described in earlier Chapters. A further comment will be given at the end of Chapter 16.

15.1. Definition of Critical Point

The diagram of state given in Fig. 1.1 shows that the liquid and gaseous states are separate only over a certain region of PVT space. At high temperatures and pressures, for example, these two states become one. In Fig. 1.1(a), the highest pressure and temperature at which the two states are separate is marked as the critical point. Figure 1.1(b), which shows pressure as a function of volume (or density), has the critical point marked as the apex of the co-existence curve. At this point, it is clear that, for the curve shown in the Figure—

$$\left(\frac{\partial P}{\partial V}\right)_T \to 0 \tag{15.1}$$

It is shown in textbooks of statistical mechanics[1] that—

$$-\frac{T}{C_V}\left(\frac{\partial P}{\partial V}\right)_T \geq 0 \tag{15.2}$$

and the critical point is defined[1] as the point where equation (15.2) reduces to an equality. Clearly equation (15.1) is an equivalent definition, provided C_V behaves in a suitable way, and in practice this restriction on C_V is found to be satisfied (see Section 15.2). The discussion of previous Chapters has been concerned with liquids near the triple point (Fig. 1.1) where the liquid and solid densities are similar. Near the critical point, however, the liquid density is close to the gaseous density, and differs markedly from that of the solid. In addition, many macroscopic properties take abnormal values in the critical region, so that the behaviour of both gas and liquid is abnormal there. For this reason, a separate discussion of critical behaviour is justified, and a brief outline will be given here.

Equation (15.1) states that at a liquid–gas critical point, a small change in pressure leads to a large change in volume. This change in volume is accomplished with only a small change in temperature, since—

$$\left(\frac{\partial T}{\partial V}\right)_P \to 0 \tag{15.3}$$

at the critical point also. This equation follows from (15.1), because—

$$\left(\frac{\partial P}{\partial V}\right)_T = -\left(\frac{\partial T}{\partial V}\right)_P \cdot \left(\frac{\partial P}{\partial T}\right)_V \tag{15.4}$$

and $(\partial P/\partial T)_V$ approaches a constant value at the critical point (Fig. 1.1). Thus at the critical point, the natural fluctuations in pressure and temperature lead to large fluctuations in density. Very close to the critical point, these fluctuations are sufficiently large that the Taylor expansion at equation (1.7b) is no longer valid, and the fluctuation theory given in earlier Chapters does not apply. However this region is small and will be excluded from the discussion, which will be concerned with the approach to the critical point from both the liquid and gaseous sides. The parameters of the critical point $(T_c, P_c$ and $\rho_c)$ will be denoted by the subscript c, and in the following the experimental behaviour of a number of physical quantities as $T \to T_c$ will be outlined. A review of experimental data in the critical region has been given by Egelstaff and Ring,[58] to which reference may be made for further details.

15.2. Compressibility and Specific Heat

The behaviour of the isothermal compressibility follows at once from equation (15.1), i.e.—

$$\chi_T = -\frac{1}{V}\left(\frac{\partial V}{\partial P}\right)_T \to \infty \tag{15.5a}$$

Also the structure factor diverges at $Q \to 0$, since—

$$[S(Q)]_{Q \to 0} = \rho k T \chi_T \to \infty \quad \text{as} \quad T \to T_c \tag{15.5b}$$

Equation (15.5b) is a consequence of the large density fluctuations discussed in the previous Section. A simple connection between χ_T and C_P is obtained from the thermodynamic formula (equation 13.3c), i.e.—

$$C_P = C_V + \chi_T T V \left(\frac{\partial P}{\partial T}\right)_V^2 \qquad (15.6a)$$

The right-hand side of this equation will diverge in the same manner as χ_T, since, as pointed out earlier, C_V and $(\partial P/\partial T)_V$ do not diverge strongly. Thus—

$$C_P \rightarrow \infty \quad \text{at the same rate as} \quad \chi_T \qquad (15.6b)$$

Experimental measurements of χ_T, $S(0)$ and C_P all show a strong divergence as the critical point is approached (e.g. Fig. 9.2). The clearest data are those for xenon,[59] which are reproduced in Fig. 15.1. These data are taken above the critical point at a density equal to ρ_c and if the divergence is represented as a power law in $|T - T_c|$ the data can be represented by—

$$\chi_T = \text{constant}\,|T - T_c|^{-\gamma} \qquad (15.7)$$

where $\gamma \gtrsim 1.1$. The theoretical significance of the value of γ will be discussed later.

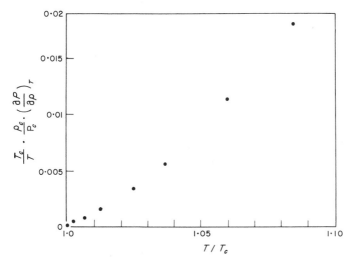

FIG. 15.1. Isothermal compressibility for xenon near T_c.[59] The non-linearity of this graph shows that γ—equation (15.7)— is greater than unity. (From Fisher,[65] Fig. 4.)

The experimental data on the specific heat at constant volume show a weak divergence at the critical point. Data on argon[60] are shown in Fig. (15.2) plotted against $\log|T - T_c|$, and the linear behaviour demonstrates that there is a divergence of approximately logarithmic form. This divergence is less pronounced than that given in equation (15.7), so confirming the assumption made in Section 15.1. The adiabatic compressibility can be related to C_V, by rewriting equation (15.6a) in the limit $C_P \gg C_V$ as—

$$C_V \simeq \chi_S T V \left(\frac{\partial P}{\partial T}\right)_V^2 \qquad (15.6c)$$

Thus χ_S should diverge in the same way as C_V, and this can be checked by measuring the velocity of sound. A logarithmic singularity in χ_S for helium[61] has been reported, for example.

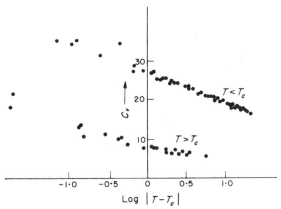

FIG. 15.2. Specific heat at constant volume of argon near the critical point, showing a logarithmic form (the units of C_V are cal/mole/°C and T is in °K). (From Bagastskii, et al.,[60] Fig. 2.)

15.3. The Co-existence Curve and Critical Isotherm

Several macroscopic properties have been discussed in Sections 15.2 and 15.3 and have been shown to diverge to infinity at the critical point. In contrast a quantity that behaves normally is the density of either the liquid or the gas. However the difference in density between the liquid and gaseous states at a given pressure and temperature vanishes at the critical point. From equation (1.5), it is seen that this difference is related to the latent heat for the transition. It has long been known[62] that the latent heat for the liquid–gas transition vanishes at the critical point roughly as $|T-T_c|^{0.4}$, but this result is not very accurate owing to experimental difficulties. However direct measurements of the liquid and gaseous densities provided more accurate data on this effect. The data on non-conducting liquids show that[58]—

$$\rho_{\text{liq}}-\rho_{\text{gas}} = \text{constant} \,|T-T_c|^{\beta} \tag{15.8}$$

where $\beta = 0.345 \pm 0.015$. Although the data on liquid metals are less satisfactory, a value of β may be deduced[58] as ~ 0.45.

Figure 15.3 is a representation of Fig. 1.1(b) showing the isotherms. The critical isotherm (i.e. for $T = T_c$) is marked and has a horizontal tangent at the critical point. If the portion of this line near the critical point is given by $|P-P_c| = \text{constant} \,|\rho-\rho_c|^{\delta}$, the value of δ that fits the data on non-conducting liquids is $\simeq 4.3$.[58]

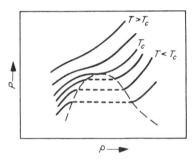

FIG. 15.3. Graph of isotherms near the gas–liquid critical point. The full lines are the isotherms and the dashed line is the boundary shown in Fig. 1.1(b).

The above results illustrate the way in which several macroscopic properties are observed to behave near the critical point. It is of some interest to explain these data theoretically, but a successful theory has not been found so far. In Section 15.5 the theory of van der Waals will be presented, which allows these properties to be discussed but does not agree with experiment. It does, however, show up the basic theoretical difficulty.

15.4. Calculation of Critical Constants

Several methods of calculating $g(r)$ were discussed in Chapter 5, and for each of these methods estimates of the equation of state may be made via equations (2.24) or (2.29). Because of the relation—

$$-\frac{1}{V}\left(\frac{\partial V}{\partial P}\right)_T = \frac{1+\rho\int_V(g(r)-1)\,\mathrm{d}r}{\rho kT} \tag{15.9}$$

the point at which the compressibility equation of state satisfies (15.1) can be evaluated from $g(r)$. For the pressure equation (2.24), a numerical evaluation of $(\partial V/\partial P)_T$ from the pressure is made. Levesque[63] has employed this method to find values of the critical parameters for argon (using the L–J potential) and a summary of his results is given in Table 15.1—here ε and σ are the parameters of the L–J potential (equation 3.10). These results confirm the superiority of the PY equation, and although the accuracy of the data is not too high, there is a reasonable measure of agreement between the theoretical and experimental values. Perhaps the least satisfactory point is the disagreement between the values forecast by the two equations of state.

TABLE 15.1

Critical parameters for argon

Critical constant	Equations for $g(r)$			Experimental value for argon	Pressure or compressibility equation
	YBG	HNC	PY		
Temperature (kT_c/ε)	$1\cdot45\pm0\cdot03$	$1\cdot25\pm0\cdot02$	$1\cdot25\pm0\cdot02$	1.26	Pressure
Density $(\rho_c\sigma^3)$	$0\cdot40\pm0\cdot05$	$0\cdot26\pm0\cdot03$	$0\cdot29\pm0\cdot03$	0·316	Pressure
Ratio $(P_c/\rho_c kT_c)$	$0\cdot44\pm0\cdot04$	$0\cdot35\pm0\cdot03$	$0\cdot30\pm0\cdot02$	0·297	Pressure
Temperature (kT_c/ε)	$1\cdot58\pm0\cdot02$	$1\cdot39\pm0\cdot02$	$1\cdot32\pm0\cdot02$	1·26	Compressibility
Density $(\rho_c\sigma^3)$	$0\cdot40\pm0\cdot03$	$0\cdot28\pm0\cdot03$	$0\cdot28\pm0\cdot03$	0·316	Compressibility
Ratio $(P_c/\rho kT_c)$	$0\cdot48\pm0\cdot03$	$0\cdot38\pm0\cdot04$	$0\cdot36\pm0\cdot02$	0·297	Compressibility

15.5. van der Waals' Equation of State

van der Waals suggested the following equation of state for the gas[1]—

$$\frac{P}{kT} = \frac{\rho}{1-b\rho} - \frac{a\rho^2}{kT} \tag{15.10}$$

and this gives isotherms similar to those shown in Fig. 15.3. For the two-phase region below the critical point the shape of this curve is abnormal,[1] but in the single-phase gas or liquid region it gives curves qualitatively similar to those observed in practice. By differentiating equation (15.10) with respect to ρ at constant temperature, it is found that the point of inflexion is given by the following equations—

$$\frac{1}{1-b\rho} + \frac{b\rho}{(1-b\rho)^2} - \frac{2a\rho}{kT} = 0 \qquad (15.11a)$$

$$\frac{2b}{(1-b\rho)^2} + \frac{2b^2\rho}{(1-b\rho)^3} - \frac{2a}{kT} = 0 \qquad (15.11b)$$

Equations (15.10) and (15.11) may be solved to find the parameters of the critical point, i.e.—

$$\left. \begin{array}{l} P_c = \dfrac{a}{27b^2} \\[2mm] kT_c = 8a/27b \\[2mm] \rho_c = 1/3b \end{array} \right\} \qquad (15.12)$$

If these expressions are employed to eliminate a and b from equation (15.10), it is found that—

$$\frac{P^*}{T^*} = \frac{8\rho^*}{3-\rho^*} - \frac{3\rho^{*2}}{T^*} \qquad (15.13)$$

where $P^* = P/P_c$, etc. Equation (15.13) suggests an independent manner of presenting PVT data, because it contains only the critical parameters. Since these parameters are known (from measurements), a number of liquids may be compared by dividing the data by the appropriate critical parameters and plotting on a non-dimensional scale. Figure 15.4 shows some data on $\rho_L - \rho_G$ for non-conducting

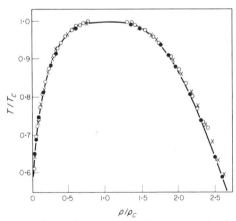

FIG. 15.4. Experimental co-existence curve for non-conducting liquids. The data have been plotted in the non-dimensional form of equation (15.13). ●, argon; ×, krypton; ○, xenon; the curve is an empirical fit for $\beta = \frac{1}{3}$. (From Guggenheim,[64] Fig. 2.)

liquids presented in this manner.[64] The fair agreement between the several sets of data verifies the "law of corresponding states", as the normalization to critical parameters is called. The curve which has been drawn through the points of Fig. 15.4 corresponds to—

$$|\rho_{\frac{1}{2}} - \rho_G| = \text{constant} \, |T - T_c|^{\frac{1}{3}} \quad (15.14)$$

which may be compared to the more accurate fit given at equation (15.8).

The index in equation (15.8) may be calculated from equation (15.13) by expanding about the critical values (i.e. unity). A comparison of this and similar results with experiment is given in Table 15.2. Unfortunately in this respect there is poor agreement between the van der Waals' theory and the data.

TABLE 15.2

Critical exponents

Exponent	Experimental value	From equation (15.13)
β	0.345 ± 0.015	0.5
γ	≥ 1.1	1.0
δ	4.3	3.0

15.6. Taylor Expansion of the Equation of State

Several improvements to the van der Waals' theory have been suggested that do not alter the predictions of Table 15.2. Fisher[65] pointed out that any theory that allowed the equation of state to be expanded as a Taylor series at the critical point would give the theoretical values at Table 15.2. In order to demonstrate this, the pressure is expanded as a series in the density, i.e.—

$$\Delta P = a(T) + b(T)\Delta\rho + c(T)(\Delta\rho)^2 + d(T)(\Delta\rho)^3 + \text{etc.} \quad (15.15)$$

where $\Delta P = P - P_c$; $\Delta\rho = \rho - \rho_c$; $\Delta T = T - T_c$, and $a(T), b(T), c(T)$ and $d(T)$ are unknown functions of temperature. The form of $a(T)$ may be found from the result $\Delta P = 0$ when $T = T_c$, so that $a(T_c) = 0$. Hence—

$$a(T) = a_1\Delta T + a_2(\Delta T)^2 + \text{etc.} \quad (15.16)$$

Also the form of $b(T)$ may be found from the result $(\Delta P/\Delta\rho) \to 0$ as

$T \to T_c$ (equation 15.1), so that $b(T_c) = 0$. Hence—

$$b(T) = b_1(\Delta T) + b_2(\Delta T)^2 + \text{etc.} \tag{15.17}$$

Now consider the pressure as a function of ρ (Fig. 15.3) for $T > T_c$. The isotherms in this region are monotonic increasing functions of ρ giving a point of inflexion at $T = T_c$, and thus as $T \to T_c$ are inconsistent with the parabolic form, or $\Delta P \propto (\Delta\rho)^2$. Hence $c(T_c) = 0$. However the shape of the critical isotherm could be consistent with a cubic form, i.e. $\Delta P \propto (\Delta\rho)^3$. Thus—

$$\left. \begin{array}{l} c(T) = c_1\Delta T + \text{etc.} \\ d(T) = d_0 + d_1\Delta T + \text{etc.} \end{array} \right\} \tag{15.17a}$$

The equations (15.16) and (15.17) are inserted into equation (15.15), and the leading terms retained to give—

$$\Delta P \simeq a_1\Delta T + b_1\Delta T \Delta\rho + d_0(\Delta\rho)^3 + \text{etc.} \tag{15.18}$$

This equation includes the leading terms in ΔT and $\Delta\rho$ and the leading cross-term, but excludes all higher terms. When $\Delta T = 0$ (i.e., on the critical isotherm) this equation gives $\delta = 3$, in agreement with Table 15.2.

Now, equation (15.18) can be rewritten as—

$$b_1\Delta T + d_0(\Delta\rho)^2 \simeq \frac{\Delta P}{\Delta\rho} - a\frac{\Delta T}{\Delta\rho}$$

$$\to 0 \quad \text{as} \quad T \to T_c \tag{15.19}$$

from equations (15.1) and (15.3). Hence—

$$\Delta\rho = \text{constant} (\Delta T)^{\frac{1}{2}} \quad \text{or} \quad \beta = \tfrac{1}{2} \tag{15.20}$$

in agreement with Table 15.2. Finally if ΔT is kept constant while small changes in pressure and density are made, equation (15.18) may be differentiated to give—

$$\left(\frac{\partial P}{\partial \rho} \right)_T \simeq b_1\Delta T + 3d_0(\Delta\rho)^2$$

or using equation (15.19)—

$$\chi_T = \frac{1}{\rho}\left(\frac{\partial \rho}{\partial P} \right)_T = \frac{\text{constant}}{\Delta T} \quad \text{or} \quad \gamma = 1 \tag{15.21}$$

Again this result is in agreement with Table 15.2, thus demonstrating that the results in the third column of the Table follow from a Taylor expansion of the equation of state.

15.7. Widom's Equation of State

The discussion of the previous Section has shown that if the equation of state can be expanded as a Taylor series near the critical point, then the predictions given in the third column of Table 15.2 are obtained. However these predictions disagree with experiment and hence a Taylor expansion is not possible. One way in which this problem has been explored is by calculating the critical behaviour of models. In this way it has been shown[65,66] that the range of the interatomic forces is a factor in determining the values of β, γ and δ, and that the van der Waals values obtained above correspond to a potential well of infinite range and of arbitrarily small depth.

Another way is through mathematical representations of the equation of state. For example, Widom[67] suggested an *ad hoc* modification to the equation of state that would allow for the experimental data. He writes the equation of state in the following special form—

$$\mu(P, T) - M(T) = \Delta\rho\Delta\tau\Phi(\Delta T, \Delta\tau) \tag{15.22}$$

where μ is the chemical potential, $M(T)$ is the chemical potential on the critical isochore (i.e. on the line $\rho = \rho_c$) in the μ, T plane, and $\Delta\tau = T - \tau(\rho)$ where $\tau(\rho)$ is the temperature at which the general isochore (at density ρ) intersects the critical isochore in the (μ, T) plane. Φ is a function that is constant in theories such as those of Sections 15.5 or 15.6, but Widom assumes that it has the form—

$$\left.\begin{aligned}
\Phi(\Delta T, \Delta\tau) &= (\Delta\tau)^{\gamma-1}\Phi\left(\frac{\Delta T}{\Delta\tau}, 1\right) \\
&= (\Delta T)^{\gamma-1}\Phi\left(1, \frac{\Delta\tau}{\Delta T}\right) \quad \text{for} \quad \Delta T > 0 \\
&= (-\Delta T)^{\gamma-1}\Phi\left(-1, -\frac{\Delta\tau}{\Delta T}\right) \quad \text{for} \quad \Delta T < 0
\end{aligned}\right\} \tag{15.23}$$

The compressibility on the coexistence curve is obtained from the result—

$$\frac{1}{\rho^2\chi_T} = \frac{1}{\rho}\left(\frac{\partial P}{\partial\rho}\right)_T = \left(\frac{\partial\mu}{\partial\rho}\right)_T = \frac{1}{\beta}\Phi(-1, 1)(\Delta\tau)^\gamma$$

$$= \frac{1}{\beta}\Phi(-1, 1)(-\Delta T)^\gamma \tag{15.24a}$$

and on the critical isochore—

$$= \Phi(0, 1)(\Delta T)^\gamma \tag{15.24b}$$

Here the coexistence curve is given by—

$$\Delta\rho = \text{constant } |\Delta\tau|^\beta \quad \text{with} \quad \Delta\tau = -\Delta T$$

and the critical isochore by $\Delta\tau = 0$ and $\Delta T > 0$. Consequently, the compressibility varies as $|\Delta T|^{-\gamma}$ near the critical point in agreement with equation (15.7). Since the function Φ permits any value of γ to be assumed, the experimental value of γ is included within this framework. It is not possible however to deduce the actual value of γ without recourse to a model of the liquid–gas system, and in this sense the model calculations are of more value than the above method.

15.8. Relationships Between the Critical Exponents

The critical-exponent theory assumes that all physical quantities approach the critical point according to a law—

$$X \propto |T - T_c|^m$$

where—

$$m = \lim_{T \to T_c} \left\{ \frac{\ln X}{|\ln|T - T_c||} \right\} \tag{15.25}$$

This example takes ΔT as the significant variable, similar definitions apply to cases where $\Delta\rho$, etc. are significant. Usually two values of m are specified $T > T_c, (m)$, and $T < T_c, (m')$. In principle, these values may differ as also may the liquid and gas values of m'. Several experimental and theoretical values of the critical exponents have been given above. There are, however, a number of relationships between the exponents, so that they are not all independent. For example on Widom's theory the critical isotherm ($\Delta T = 0$, characterized by the exponent δ) may be derived from equation (15.22), i.e.—

$$\mu - \mu_c = \Delta\rho\Phi(0, 1)(\Delta\tau)^\gamma \tag{15.26}$$

where $\mu_c = M(T_c)$ is the chemical potential at the critical point. On the co-existence curve—

$$\Delta\tau = \text{constant} \, (\Delta\rho)^{1/\beta}$$

so that equation (15.26) becomes—

$$\Delta\mu = \text{constant} \, (\Delta\rho)^{1 + \gamma/\beta}\Phi(0, 1) \tag{15.27}$$

But at constant temperature $\Delta\mu = \Delta P/\rho$, so that the required exponent is—

$$\delta = 1 + \frac{\gamma}{\beta} \tag{15.28}$$

This derivation demonstrates the value of the mathematical approach in suggesting such useful results. The relation (15.28) is exactly satisfied by the van der Waal's theory ($\beta = \frac{1}{2}, \gamma = 1, \delta = 3$) and roughly satisfied by the experimental values ($\delta \simeq 4\cdot3$; $\gamma \simeq 1\cdot1$, $\beta \simeq 0\cdot33$). However, since the above theory is approximate, it is possible that the relation (15.28) may not be satisfied in all cases, and a more general relation is sometimes postulated[58]—

$$\gamma' \geq \beta(\delta - 1) \tag{15.29}$$

A listing of the critical exponents, their experimental values and the relationships between them is given in the literature.[58]

15.9. Behaviour of $S(Q)$ near the Critical Point

Equation (15.5b) shows that $S(0)$ is proportional to $|T - T_c|^{-\gamma}$ near the critical point. Since the density fluctuations leading to this behaviour are very long range, it is expected that they will become less pronounced at short range. In this case $S(Q)$ will decrease with increasing Q possibly as shown in Fig. (15.5a). As the temperature is lowered away from T_c, the shape of $S(Q)$ takes on the forms shown successively in Fig. 15.5(b), (c) and (d).

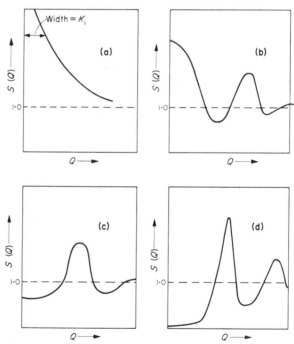

Fig. 15.5. Qualitative behaviour of $S(Q)$ at several temperatures between the critical point and the triple point: (a), critical point; (b), high temperature; (c), average temperature; (d), low temperature.

Ornstein and Zernike have calculated the form of $S(Q)$ at low Q, assuming that the correlation functions can be Taylor-expanded at the critical point. In view of the limitations of Taylor expansions mentioned, above this theory will not be accurate but will give the general behaviour only. The starting point is equation (5.9), i.e.—

$$h(\mathbf{r}) = c(\mathbf{r}) + \rho \int_V c(|\mathbf{r} - \mathbf{r}'|)h(\mathbf{r}') \, d\mathbf{r}' \qquad (15.30)$$

Since $h(r)$ is long ranged as $T \to T_c$, it is difficult to evaluate directly, but equation (15.30) may be reduced to an equation in $h(r)$ if it is assumed that $c(r)$ is short ranged. Thus for large r assume $c(r) \to 0$ and may be neglected in equation (15.30). Also the factor $c|\mathbf{r} - \mathbf{r}'|$ may be expanded about $\mathbf{r}$, since for non-zero values of c the quantity $R = |\mathbf{r} - \mathbf{r}'|$ is small. After these steps, and the further assumption that all moments of $c(r)$ are bounded, equation (15.30) becomes—

$$h(r) \simeq \rho C h(r) + \rho C \sigma^2 \nabla^2 h(r) \qquad (15.31a)$$

where—

$$C = \int_V c(r) \, d\mathbf{r}$$

and—

$$\sigma^2 = \frac{1}{6C} \int_V r^2 c(r) \, d\mathbf{r} \qquad (15.31b)$$

If $c(r)$ is short ranged compared to $h(r)$, then it should be found that—

$$\sigma^2 \ll s^2 \qquad (15.32a)$$

where—

$$s^2 = \frac{1}{6H} \int_V r^2 h(r) \, d\mathbf{r} \qquad (15.32b)$$

and $H = \int_V h(r) \, d\mathbf{r}$. Now the solution of equation (15.31a) that satisfies equation (15.32a) is—

$$h(r) = \frac{H}{s^2} \frac{e^{-r/s}}{4\pi r} \qquad (15.33a)$$

and from equations (15.31b) and (15.30)—

$$s^2 = \frac{\sigma^2 C}{1 - C} = \sigma^2 H \qquad (15.33b)$$

The latter result confirms the assumption (15.32a). Thus this method gives a self-consistent solution, but it is not necessarily the correct solution.

Because $H \to |T - T_c|^{-\gamma}$ as $T \to T_c$ this result gives $s^2 \propto |T - T_c|^{-\gamma}$ near the critical point. Equation (15.33a) may be Fourier-transformed to give—

$$S(Q) \simeq \rho k T \chi_T \frac{K_1^2}{K_1^2 + Q^2} \qquad (15.34)$$

where $K_1 = 1/s$. This expression is a Lorentzian curve whose width vanishes with an exponent $\nu = \gamma/2$ as $T \to T_c$. Such behaviour has been observed in many experiments,[58] although the value of the exponent is not known precisely. Although the experimental behaviour is similar to equation (15.34), this does not necessarily confirm the above theory. Experiments of high accuracy may be expected to show deviations related to the non-analytic behaviour of macroscopic properties near the critical point.

The qualitative behaviour of $S(Q)$ as a function of temperature is shown in Fig. 15.5. Near the critical point, equation (15.34) is obeyed approximately and $S(Q)$ is dominated by the strong peak at the origin. The oscillations in $S(Q)$ are likely to be observable, but do not affect the main feature of the curve. At a lower temperature (in the liquid phase) the peak at $Q \sim 0$ is still a major feature, but the oscillatory features are relatively more significant. Typical liquid data are shown in Fig. 15.5(c) and (d) for comparison, Fig. 15.5(d) applying near the triple point, whereas Fig. 15.5(c) applies at a somewhat higher temperature where the compressibility is greater.

15.10. Behaviour of $S(Q, \omega)$ near the Critical Point

The frequency dependence of $S(Q, \omega)$ near the critical point may be obtained from the visco-elastic theory by taking appropriate limits. In particular, the viscous terms in equation (13.27) are ignored, since to a first approximation they behave normally as $\chi_T \to \infty$ at T_c. The dispersion law for low values of Q and low frequencies is given in equation (13.9). In the present case the $C_P \gg C_V$ limit of this equation is required for which $\omega \sim D_T' Q^2 \ll D_T Q^2$ where—

$$D_T' = \frac{\lambda}{C_P \rho} \qquad (15.35)$$

Thus equations (13.27) and (13.30) become—

$$\omega^2 = \frac{q^2}{M \rho \chi_T} \left[1 - \frac{C_P}{C_V} \frac{i\omega}{\omega_0 - i\omega} \right]$$

or—

$$\frac{q^2}{M\rho\chi_T} \simeq \frac{i\omega q^2}{M\rho\chi_T D_T'} \tag{15.36}$$

giving a non-propagating mode of lifetime $1/D_T'q^2$. Comparison of equations (15.36) and (14.27) shows that in this case—

$$V(Q, \omega) \simeq -1/\rho\chi_T D_T'$$

and the scattering law (equation 14.26) becomes, in the low-frequency limit—

$$S(Q, \omega) \simeq \frac{S(Q)}{\pi} \cdot \frac{D_T'Q^2}{(D_T'Q^2)^2 + \omega^2} \tag{15.37}$$

This result was first obtained by van Hove. Again, this is a Lorentzian curve and its width is $D_T'Q^2$. Since the width of the small Q peak of $S(Q)$ (from equation 15.34) is $\sim 1/s^2$ and $C_P \propto \chi_T$, it can be shown that—

$$[D_T'Q^2]_{\text{average}} \propto |T - T_c|^{2\gamma} \tag{15.38}$$

Thus the frequency width of $S(Q, \omega)$ vanishes very sharply as $T \to T_c$, and this has been confirmed in a number of experiments.[58] The precise value of the exponent is unknown. Equations (15.34) and (15.37) illustrate that at the critical point the thermal fluctuation component of $S(Q, \omega)$ becomes larger than the other components, and gives rise to a sharp peak for $Q \sim 0$ and $\omega \sim 0$. If the values of Q and ω are increased, the effect of this peak becomes less important, confirming that the critical point is a macroscopic or infinite volume phenomenon.

Symbols for Chapter 15

$a(T)$		γ	Critical exponent for compressibility
$b(T)$	Coefficients in Taylor expansion		
$c(T)$	of equation of state	δ	Critical exponent for critical isotherm
$d(T)$			
a	Coefficients in van der Waals	η	Critical exponent for $h(r)$
b	equation of state	ν	Critical exponent for $K_1 = 1/s$
c	Subscript indicating critical value	σ^2	Mean-square value of r in $c(r)$
$M(T)$	Chemical potential on the critical isochore	Φ	Homogeneous function having properties given in equation (15.23)
m	Critical exponent (arbitrary)		
s^2	Mean square value of r in $h(r)$		
β	Critical exponent for coexistence curve.		

CHAPTER 16

Quantum Liquids

16.1. Comparison of Classical and Quantum Liquids

All the previous discussion has been concerned with classical liquids, that is liquids at relatively high temperatures and usually of high atomic mass. For such liquids (Table 9.1) the wavelength of an atom is small compared to the interatomic distances and each atom may be thought of as a distinct particle. In contrast, the atom in a quantum liquid has a wavelength comparable to the interatomic spacing and it is no longer possible to discuss individual atoms, but only the atomic density.

Another important feature of quantum liquids is the statistical system to which the atoms belong. A system of identical particles will satisfy Bose statistics if its wave function is symmetric under an interchange of any pair of particle co-ordinates, but if the wave function is anti-symmetric, the system satisfies Fermi statistics. It is shown in standard texts[3,4] that for a Bose system any number of particles can have the momentum $\hbar q$ in a given state of the system, and that the average number $\langle n_q \rangle$ is—

$$\langle n_q \rangle = \frac{z}{\exp(E_q/kT) - z} \qquad (16.1)$$

where z is the fugacity, and E_q is the average energy for the state of momentum $\hbar q$. The corresponding quantity in Fermi statistics where either zero or one particle can be in a state of momentum $\hbar q$, is—

$$\langle n_q \rangle = \frac{z}{\exp(E_q/kT) + z} \qquad (16.2)$$

At high temperatures both equations (16.1) and (16.2) reduce to the equation of classical statistics or the Boltzman statistics, i.e.—

$$\langle n_q \rangle = z \exp\left\{-\frac{E_q}{kT}\right\} \qquad (16.3)$$

which has been used in earlier Chapters.

Thus quantum liquids are of two kinds—Bose and Fermi—and the basic examples are liquid ^{4}He and liquid ^{3}He, respectively. This Chapter will be devoted to a brief outline of some of the properties of these liquids in order to illustrate some of the new features that are apparent in quantum liquids. For a detailed discussion, the reader is referred to one of the standard texts.[68] A large part of the present discussion will be concerned with liquid ^{4}He.

16.2. Wave Function for Quantum Liquids with Short-range Forces

(The discussion of this Section follows that of Jastrow.[69])

In this Section the ground-state properties of a quantum liquid are discussed. It will be assumed that there are N atoms in a volume V, and that they interact through central two body forces represented by the potential $u(r)$. The Hamiltonian of the system is—

$$H = -\frac{\hbar^2}{2M} \sum_{i=1}^{N} \nabla_i^2 + \sum_{i<j=1}^{N} u(r_{ij}) \tag{16.4}$$

This Hamiltonian is the quantum mechanical counterpart of the classical Hamiltonian given in equations (2.1) and (2.17). If it is assumed that the range (σ) of the core of the potential $u(r)$ is short compared to the interparticle separation [for example $u(r)$ is given by the hard-sphere potential equation (3.12)], then the eigenfunctions of equation (16.4) have the form—

$$\psi = S(\mathbf{r}_1 \ldots \mathbf{r}_n) \prod_{i<j=1}^{N} F(r_{ij}) \tag{16.5}$$

where $F(r_{ij})$ vanishes for $r < \sigma$ and approaches unity for $r \gg \sigma$. For a Bose system, S is unity, but for a Fermi system S is a Slater determinant of one particle wave functions.

The expectation value of H in a Bose system is calculated from—

$$\langle H \rangle = \frac{\int_V \cdots \int \prod_{i<j=1}^{N} F^*(r_{ij}) H \prod_{i<j=1}^{N} F(r_{ij}) \, d\{N\}}{\int_V \cdots \int \prod_{i<j=1}^{N} |F(r_{ij})|^2 \, d\{N\}} \tag{16.6}$$

and substituting H from equation (16.4), the cross-terms in the kinetic energy vanish upon integration, and—

$$\langle H \rangle = N(N-1)$$

$$\frac{\int e^{-\frac{1}{2}\Sigma w^*(r_{ij})} \left[-\frac{\hbar^2}{2M} \frac{\nabla_1^2 F(r_{12})}{F(r_{12})} + \frac{1}{2} u(r_{12}) \right] e^{-\frac{1}{2}\Sigma w(r_{ij})} \, d\{N-2\}}{\int |e^{-\frac{1}{2}\Sigma w(r_{ij})}|^2 \, d\{N\}} \tag{16.7}$$

where—

$$e^{-w(r_{ij})/2} = F(r_{ij}) \qquad (16.8)$$

A comparison of equation (16.7) and equations (2.15) and (2.16) shows that the classical limit is obtained from equation (16.7) if—

$$w(r_{ij}) \equiv \frac{u(r_{ij})_{Cl}}{kT} \qquad (16.9)$$

Thus in a theory of this type, there will be quantum mechanical formulae corresponding to each of the classical formulae used earlier, and this may be noted by stating that f_{ij} in equation (2.30) is equivalent to $F^2(r_{ij}) - 1$.

For example, the pair distribution function is defined by the integration of ψ^2 over atoms $3 \ldots N$, i.e.—

$$g(r_{1\,2}) = \frac{N(N-1) \int_V \ldots \int |\exp -\tfrac{1}{2}\Sigma w(r_{ij})|^2 \, d\{N-2\}}{\rho^2 \int_V \ldots \int |\exp -\tfrac{1}{2}\Sigma w(r_{ij})|^2 \, d\{N\}} \qquad (16.10)$$

which is analogous to equation (2.2) with $n = 2$. Also for a Bose system the kinetic ($\overline{K}$) and potential ($\overline{U}$) energy may be written[69]—

$$\left. \begin{aligned} \frac{\overline{U}}{N} &= \frac{\rho}{2} \int_V u(r)g(r) \, dr \\ \frac{\overline{K}}{N} &= \frac{\rho \hbar^2}{8M} \int_V \frac{\partial g(r)}{\partial r} \frac{\partial w(r)}{\partial r} \, dr \end{aligned} \right\} \qquad (16.11a)$$

and the pressure is—

$$P = \frac{2}{3}\frac{\overline{K}}{N} - \frac{\rho^2}{6} \int_V g(r)r \frac{\partial u(r)}{\partial r} \, dr \qquad (16.11b)$$

The potential-energy equation is of the same form as equation (2.18), but the kinetic energy is greater than $\tfrac{3}{2}NkT$. For a Fermi system, the expressions for the kinetic and potential energies are more complex[69] than for the Bose system, and will not be reproduced here.

If equation (16.5) is a good representation of the ground-state wave function, then all the properties of a quantum liquid may be calculated from pair correlation theory as shown below.

16.3. Calculation of $g(r)$ for a Bose Liquid

The general definition of an n particle distribution function is a natural extension of equation (16.10), namely—

$$g_n(\mathbf{r} \dots \mathbf{r}_n) = \frac{N!}{\rho^n(N-n)!} \frac{\int |\psi|^2 \, d\{N-n\}}{\int |\psi|^2 \, d\{N\}} \qquad (16.12)$$

Arguments analogous to those used in Chapter 5 will show that g_2 and g_3 satisfy—

$$\nabla_1 g(r_{12}) = -g(r)\nabla_1 w(r_{12}) - \rho \int_V g_3(\mathbf{r}_1, \mathbf{r}_2, \mathbf{r}_3)\nabla_1 w(\mathbf{r}_{31}) \, d\mathbf{r} \qquad (16.13)$$

which is the analogue of equation (5.5).

Wu and Feenburg[70] have used the superposition approximation (equation 5.7) in equation (16.13) in order to evaluate $w(r)$ from an experimental $g(r)$. Their result is shown in Fig. 16.1(a), from which it can be seen that in this approximation $w(r)$ is positive for all reasonable values of r. Massey[71] uses a set of trial functions for $g(r)$ and then calculates $w(r)$ from equation (16.13) and the superposition approximation. Finally he calculates $\langle H \rangle$ and finds which $g(r)$ gives a minimum value. This result for $g(r)$ is compared to experiment in Fig. 16.1(b); the agreement is only fair.

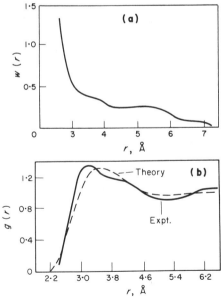

FIG. 16.1. (a) Component of ground-state wave function calculated from YBG equation (from Wu and Feenberg,[70] Fig. 2); (b) Pair distribution $g(r)$ calculated from YBG equation (from Massey,[71] Fig. 6).

It is believed that the wave function (equation 16.5) gives a good description for the ground state of liquid ^{4}He, but many of the difficulties described in Chapters 5 and 7 must be overcome before this can lead, through equation (16.13), to the successful calculation of $w(r)$ and $g(r)$.

16.4. Co-operative Modes of Motion in ^{4}He near $T = 0$

As $T \to 0$, a Bose liquid will fall into its ground state and, in the limit of non-interacting atoms, all atoms will have the same momenta. This state will be separated from higher states by an energy gap, ensuring that it is a well defined state for the liquid. As will be discussed later, it is a state of vanishing viscosity, and hence the liquid is called a "superfluid". The superfluid has many interesting properties, which can be understood from a discussion of the co-operative modes of motion.

In the case of ^{4}He, the liquid passes from the normal to the superfluid state at a temperature of 2·2°K. The curve of specific heat as a function of temperature is shown in Fig. 16.2(a) and has a "λ" shape giving rise to the name λ transition or λ point. Actually, a curve of this shape is characteristic of a specific heat varying as $\log|T - T_\lambda|$. At low temperatures, the specific heat varies as T^3 (Fig. 16.2b), showing that the low-energy excitations are long wavelength (Debye) phonons. If it is assumed that as $T \to 0$, these and other elementary excitations have a long lifetime, then it is possible to calculate the relation between frequency and wave number from equation (9.12). Because a given value of $Q = Q_0$ corresponds to one mode of frequency ω_0, the function $S(Q, \omega)$ will consist of sharp δ function like peaks at $\omega = \pm\omega_0$. Thus write—

$$S(Q, \omega) = A\delta(\omega \pm \omega_0)\, e^{\hbar\omega/2kT} \tag{16.14}$$

where A is a constant defined by—

$$S(Q_0) = \int_{-\infty}^{+\infty} S(Q_0, \omega)\, d\omega = 2A \cosh \frac{\hbar\omega_0}{2kT} \tag{16.15a}$$

Hence equation (9.12) can be written—

$$\frac{\hbar Q_0^2}{2M} = \int_{-\infty}^{+\infty} \omega S(Q_0, \omega)\, d\omega = 2A\omega_0 \sinh \frac{\hbar\omega_0}{2kT} \tag{16.15b}$$

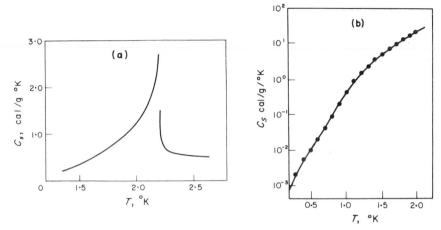

FIG. 16.2. Specific heat of liquid ^{4}He along the saturated-vapour-pressure curve
(C_s). Curve (a) shows the behaviour near the λ transition at $2\cdot2°$K (W. H. Keesom,
"Helium," Elsevier). Curve (b) shows the experimental data ($\bullet$) compared to the values
calculated (—) from Fig. 16.3 (P. J. Bendt, R. D. Cowan, and J. L. Yarnell, *Phys. Rev.*,
1959, **113**, 1386).

The combination of equation (16.15a) and (b) yields—

$$Q_0 = \sqrt{\frac{2M\omega_0 S(Q)}{\hbar}} \tanh \frac{\hbar\omega_0}{2kT} \xrightarrow[T\to0]{} \sqrt{\frac{2M\omega_0 S(Q)}{\hbar}} \qquad (16.16)$$

which is the well known result of Feynman.[73] This equation shows
that for a low value of Q where the frequency–wave-number relationship is $\omega_0 = cQ_0$—

$$[S(Q_0)]_{Q_0\to0} \to \frac{\hbar\omega_0}{2Mc^2} = \frac{\hbar Q_0}{2Mc} \qquad (16.17)$$

unlike the classical result (equation 6.23).

At higher values of Q, equation (16.16) leads to an $(\omega–Q)$ relationship
of the type given in Fig. 16.3: the experimental $S(Q)$ is used in equation
(16.16), since the calculations (Section 16.3) are not yet reliable.
Experimental measurements confirm that $S(Q, \omega)$ has approximately
the form given in equation (16.14), but there is another term, due to
higher-order effects, that gives a broad spectrum. This term modifies
equation (16.16), and hence a graph of the observed $(\omega–Q)$ values of
the peaks in $S(Q, \omega)$ differs from that calculated from equation (16.16).
The observed values are also plotted in Fig. 16.3. There is a qualitative
correspondence between the two curves, which confirms the existence
of an energy gap (Δ) for short-wavelength excitations, but there are

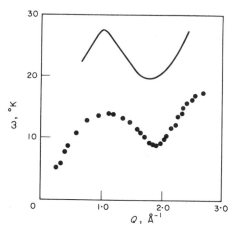

FIG. 16.3. Graph of peak positions in $S(Q, \omega)$. The full line is equation (16.16), and the dots are the experimental data.[72]

significant quantitative differences showing the importance of the higher order contributions to $S(Q)$.

It is important to show that the data given in Fig. 16.3 include all the elementary excitations of the superfluid [for this step the reader is referred to the literature[68] on quantum liquids], but if this is assumed, it is possible to calculate the thermodynamic properties of the superfluid from the $(\omega$–$Q)$ data. Calculations of this kind are in good agreement with the measured values of C_s, etc. (as shown in Fig. 16.2b), so confirming the underlying theoretical assumptions concerning liquid He II.

16.5. Critical Velocity for Superfluid

It is interesting to consider the behaviour of the $(\omega$–$Q)$ relation as a liquid is taken from the high-temperature critical region through a normal liquid region to a superfluid. The qualitative behaviour is shown in Fig. 16.4. At the critical region, long-wavelength sound waves are not propagated, but high-frequency waves may be. In a normal liquid both low- and high-frequency waves are propagated, but for a Q value equal to the peak of $S(Q)$ [Fig. 6.2] the velocity vanishes. A superfluid differs from a normal liquid in that modes of this wave number propagate (i.e. the velocity is non vanishing). Landau has explained, qualitatively, how this effect gives rise to superfluidity, i.e. absence of viscosity.

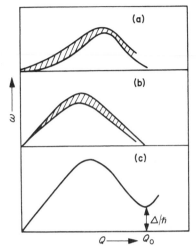

FIG. 16.4. Qualitative plot of $(\omega-Q)$ curves for : (a), critical point ; (b), normal liquid ; (c), superfluid. The shaded areas indicate regions where the modes have short life times.

Assume a body is moving through the liquid (at $T \to 0$) with velocity $\mathbf{v}_i$ and momentum $\mathbf{p}_i$, and that it looses an energy δE by creation of elementary excitations allowing its momentum to become $\mathbf{p}_f$. Then—

$$\delta E = \frac{\mathbf{p}_i^2}{2M} - \frac{\mathbf{p}_f^2}{2M} = \mathbf{v}_i \, \delta \mathbf{p} - \frac{(\delta p)^2}{2M} \qquad (16.18)$$

where $\delta \mathbf{p} = \mathbf{p}_i - \mathbf{p}_f$. In order to create an excitation therefore the body must have a velocity greater than—

$$v_i > \frac{\delta E}{\delta p} \qquad (16.19a)$$

Since there is no resistance to motion (i.e. zero viscosity) if no excitations are created, it is clear that a necessary condition for superfluidity is—

$$\frac{\delta E}{\delta p} > 0 \qquad (16.19b)$$

where δE and δp are the energy and momentum of an elementary excitation respectively. For $Q \to 0$, $\delta E/\delta p \sim c$, so that only large values of Q need be considered. If the $(\omega-Q)$ curve has the form shown in Fig. 16.4(b), then $\delta E/\delta p = 0$ for $Q = Q_0$, and viscous resistance certainly occurs. In contrast, if the $(\omega\text{-}Q)$ relation has the form shown in Fig. 16.4(c) for the superfluid, then the minimum value of $\delta E/\delta p$ is $\Delta/\hbar Q_0$. Thus, on this picture the liquid will be superfluid for velocities less than $\Delta/\hbar Q_0$.

It is found in practice that the critical velocity for the loss of superfluidity is ill defined, but is of the order of several centimetres/sec or two orders of magnitude less than 80 metres/sec calculated on the above model. The reason for this discrepancy was pointed out by Feynman[73] who suggested that there are other excitations at higher Q values that give a low value of $\delta E/\delta p$. These excitations are "vortex lines". It is imagined that liquid flows out of an orifice of width (d) creating many vortices spaced at a distance (x) apart. If (a) is the radius of the hole in the centre of the vortex, then the energy per unit length is—

$$\varepsilon = \int_a^d \frac{1}{2} \cdot 2\pi r \, \mathrm{d}r \rho \left(\frac{\hbar}{Mr}\right)^2 = \pi \rho \left(\frac{\hbar}{M}\right)^2 \log \frac{d}{a} \qquad (16.20)$$

Now the number of lines per unit length can be calculated from Stokes theorem by taking a line integral along the direction of flow for unit distance and returning outside the stream, i.e. the number of lines per unit length is—

$$\frac{1}{x} = \frac{vM}{2\pi\hbar}$$

Since v/x lines/sec are created, the total energy absorbed by the vortex lines is—

$$\frac{v^2 M}{2\hbar} \cdot \rho \left(\frac{\hbar}{M}\right)^2 \log \frac{d}{a} \qquad (16.21a)$$

This is equal to the available kinetic energy, or—

$$vd\frac{M\rho v^2}{2} \qquad (16.21b)$$

Equations 16.21(a) and (b) give the minimum velocity for creation of vortex lines as—

$$v \sim \frac{\hbar}{Md} \log \frac{d}{a} \qquad (16.22)$$

Some problems arise in the choice of values of d and a, but reasonable values are $d \sim 10^{-5}$ cm and $\log d/a \sim 6$, giving $v \sim 100$ cm/sec. This value is nearly two orders of magnitude smaller than Landau's value, and demonstrates that the Landau condition, while being necessary, is not by itself sufficient for superfluidity.

16.6. The Two-fluid Model of He II

The properties of liquid helium II were explained, historically, on the basis of a two-fluid model. It is postulated that below $2 \cdot 2°K$ helium consists of a "normal fluid" and a superfluid, and the dynamics of this system are discussed by an extension to the hydrodynamic theory of Chapter 13. At absolute zero, the liquid is all superfluid, and at $2 \cdot 2°K$ it is all normal liquid, while in between there are varying proportions of the two components. From the point of view of the Bose liquid this model is physically meaningless, but has a valuable mathematical content. At absolute zero, the liquid is in the ground state and is a true superfluid. However at finite temperatures, some elementary modes are excited and the liquid has a finite viscosity. On the two-fluid model, this is interpreted as the existence of a finite density of normal fluid having a normal viscosity. Thus, it can be seen that the density of "normal" fluid is just given by the number of modes excited at any temperature. It can be shown[3] that a convenient definition of the density (ρ_n) of "normal" fluid is—

$$\rho_n = \frac{\hbar}{(2\pi)^3 u^2 M} \int d\mathbf{q} \; \mathbf{u} . \mathbf{q} \langle n_q \rangle \tag{16.23}$$

where $\langle n_q \rangle$ is the average occupation number for modes of wave number q, and $\mathbf{u}$ is the average group velocity of an excitation with respect to the superfluid—

$$\mathbf{u} = \frac{\int d\mathbf{q} \langle n_q \rangle \nabla_q \omega_q}{\int d\mathbf{q} \langle n_q \rangle} \tag{16.24}$$

If the excitations are very weakly interacting, they can be treated as a liquid of density ρ_n, and so explain the success of the two-fluid model.

16.7. Wave Propagation in a Superfluid

The usual conservation equations, in a linearized form, are—

Continuity

$$M\frac{\partial \rho}{\partial t} + \nabla . \mathbf{p} = 0 \tag{16.25a}$$

Momentum

$$\frac{\partial \mathbf{p}}{\partial t} + \nabla P = 0 \tag{16.25b}$$

where P is the pressure, $\rho = \rho_n + \rho_s$ and the momentum density is—

$$\mathbf{p} = M\rho_s\mathbf{v}_s + M\rho_n\mathbf{v}_n \tag{16.25c}$$

In this equation, $\mathbf{v}_s$ and $\mathbf{v}_n$ are the velocities of the normal and super-fluid, respectively, which are unknown. To eliminate these velocities, the linearized equations for conservation of entropy and energy are introduced,[3] i.e.—

$$\frac{\partial S}{\partial t} + \nabla \cdot \mathbf{j}_s = 0 \tag{16.25d}$$

$$\frac{\partial \mathbf{p}'}{\partial t} + S\nabla T = 0 \tag{16.25e}$$

where S is the entropy per unit volume, $\mathbf{j}_s = S\mathbf{v}_n$ is the entropy current and $\mathbf{p}' = M\rho_n(\mathbf{v}_n - \mathbf{v}_s)$ is the momentum of relative motion between the normal and superfluid. These equations represent important physical statements, the first that entropy is carried only by the normal component, and the second that a temperature gradient causes a separation of normal and superfluid (fountain effect).

It is now possible to eliminate $\mathbf{v}_s$ and $\mathbf{v}_n$ from equations (16.25) to obtain, after some algebra—

$$M\frac{\partial^2 \rho}{\partial t^2} - \nabla^2 P = 0 \tag{16.26a}$$

$$M\frac{\partial^2 \rho}{\partial t^2} - \frac{M\rho}{S}\frac{\partial^2 S}{\partial t^2} + \frac{\rho_s}{\rho_n}S\nabla^2 T = 0 \tag{16.26b}$$

Finally by choosing T and ρ to be independent parameters, S and P can be eliminated from equation (16.26) by the equations—

$$\left.\begin{array}{l} \dfrac{\partial^2 S}{\partial t^2} = \left(\dfrac{\partial S}{\partial T}\right)_\rho \dfrac{\partial^2 T}{\partial t^2} + \left(\dfrac{\partial S}{\partial \rho}\right)_T \dfrac{\partial^2 \rho}{\partial t^2} \\[3mm] \nabla^2 P = \left(\dfrac{\partial P}{\partial T}\right)_\rho \nabla^2 T + \left(\dfrac{\partial P}{\partial \rho}\right)_T \nabla^2 \rho \end{array}\right\} \tag{16.27}$$

In general, this will give two coupled equations for T and ρ, but in the limit $T \rightarrow 0$, equations (16.26) reduce to—

$$\nabla^2 \rho - M\left(\frac{\partial \rho}{\partial P}\right)_T \frac{\partial^2 \rho}{\partial t^2} = 0 \tag{16.28a}$$

$$\nabla^2 T - \left[\frac{M\rho\rho_n}{S^2\rho_s}\left(\frac{\partial S}{\partial T}\right)_\rho\right]\frac{\partial^2 T}{\partial t^2} = 0 \tag{16.28b}$$

These equations describe, respectively, the propagation of density (sound) waves and temperature (or entropy) waves. At absolute zero, the velocities are in the ratio $\sqrt{3}$, but the ratio is larger at higher temperatures. Hence a new feature of the superfluid is the propagation of heat pulses (second sound), which has been confirmed experimentally in good agreement with the above theory.[68]

These thermal waves can be thought of as compression waves in a non-interacting phonon gas. Compared to the normal hydrodynamic model, they replace the non-propagating thermal fluctuations. Consequently the behaviour of the thermal fluctuations reflects the state of the liquid, and it is interesting to follow the qualitative changes that occur on passing from the critical region through to the superfluid region. This may be done, most conveniently, by sketching the

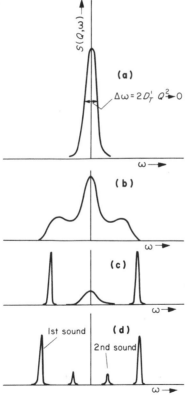

FIG. 16.5. Qualitative picture of $S(Q, \omega)$ for low Q values at temperatures between the critical point and the superfluid region: (a), critical point; (b), high-temperature liquid; (c), normal liquid; (d), superfluid.

scattering law $S(Q, \omega)$ as a function of ω for a chosen very small value of Q. Figure 16.5 (a) is a sketch of this kind showing the thermal peak dominant at the critical point due to the strong coupling between entropy and density fluctuations. At temperatures below the critical point the sound-wave peaks appear as wings on the central peak, and become sharper as the central peak decreases with decreasing temperature. In a liquid region, such as that marked in Fig. 1.1, $S(Q, \omega)$ has the form shown at Fig. 16.5(c) with intense and sharp sound-wave peaks, but a small fairly broad thermal peak. For liquid helium near the λ point, the sound-wave velocity is reduced and the sound absorption increases, and after passing through this region, two types of wave propagation become possible for $T < T_\lambda$. In the superfluid phase the thermal peak becomes the propagating modes of second sound as shown in Fig. 16.5(d), but owing to the weak coupling between entropy and density fluctuations at low temperatures the intensity is very low.

16.8. Comment on Critical, Normal and Quantum Liquids

The properties of critical and quantum liquids have been discussed briefly and the theoretical background has been sketched in a comparative style to that employed for "normal" liquids. This has served to indicate the very wide range of properties it is possible for a liquid to take on: the compressibility and viscosity may be quoted as examples of properties that may take on almost any value.

It has been seen that a relatively simple theory can be used to discuss these properties in an overall manner, although its validity has not been examined properly. For example, the significance of the triplet correlation function has not been discussed even though it is likely to be important in some regions, particularly near the critical point. This parallels the omission of many of the important criticisms of the pair theory for metals (e.g. can the energy really be described in pseudo-potential theory?) and the omission of a large part of many body theory. However it has been comforting to find that an outline of the whole of the liquid state can be given, and that from the general body of knowledge a theory of liquids is beginning to emerge. Thus although the liquid state covers only a narrow portion of the PVT diagram, it includes an extensive range of values of the physical properties and a general theoretical treatment would be needed to cover them all. It has been shown that through various approximations, the experimentally observed behaviour may be understood for all these types of liquids, but that much of the detail remains a mystery.

Symbols for Chapter 16

a Radius of "hole" in vortex lines

d Width of orifice (Section 16.6)

E_q Energy of particle or excitation of momentum $\hbar q$

$F(\mathbf{r}_{ij})$ Pair factor in ground-state wave function (equation 16.5)

$\mathbf{j}_s$ Entropy current

$\overline{K}$ Kinetic energy of atom in quantum liquid

n_q Number of particles or excitations of momentum $\hbar q$

Q_0 Wave number of elementary excitation

$\mathbf{p}$ Momentum density (including $\mathbf{p}$ subscript or prime)

$\mathbf{u}$ Average group velocity of an excitation

$\mathbf{v}_s$ Velocity of element of superfluid

$\mathbf{v}_n$ Velocity of element of normal fluid

$w(\mathbf{r}_{ij})$ $-2 \log F(\mathbf{r}_{ij})$

x Distance between vortex lines

Δ Energy gap in spectrum of elementary excitations

ρ_n Density of normal fluid

ρ_s Density of superfluid

ψ Wave function for ground-state of quantum liquid (equation 16.5)

ω_0 Frequency of elementary excitations

References

1. Landau, L. D., and Lifshitz, E. M., "Statistical Physics." Pergamon, London. 1958.
2. Yule, G. U., and Kendall, M. G., "An Introduction to the Theory of Statistics," Chapter 10. Charles Griffin, London. 1950.
3. Huang, K., "Statistical Mechanics." Wiley, New York. 1965.
4. Hill, T. L., "Statistical Mechanics." McGraw-Hill, New York. 1956.
5. Rice, S. A., and Gray, P., "The Statistical Mechanics of Simple Liquids." Wiley, New York. 1965.
6. Schiff, L., "Quantum Mechanics." McGraw-Hill, New York. 1949.
7. Mott, H., and Massey, H., "Theory of Atomic Collisions." Oxford University Press, London. 1947.
8. Bernal, J. D., *Nature, Lond.*, 1960, **185**, 68.
9. Ruelle, D., *In* "The Equilibrium Theory of Classical Fluids," (Ed. H. L. Frisch and J. L. Lebowitz), p. 1.1. Benjamin, New York. 1964.
10. Hirschfelder, J. O., Curtis, C. F., and Bird, R. B., "Molecular Theory of Gases and Liquids." Wiley, New York. 1954.
11. Kennard, E. H., "Kinetic Theory of Gases." McGraw-Hill, New York. 1938.
12. McDowell, M. R. C., (Editor) "Proceedings of the 3rd International Conference of Physics of Electronic and Atomic Collisions." North-Holland, Amsterdam. 1965. [Figure 3.3a is taken from E. Rothe *et al.*, p. 927, Fig. 3.3b & c from R. B. Bernstein *et al.*, p. 895 and Fig. 3.4 from I. Amdur *et al.*, p. 934.]
13. Harrison, W. A., "Pseudo-potentials in the Theory of Metals." Benjamin, New York. 1966. [A simple review of this subject has been given by Ziman, J. M., *Adv. Phys.*, 1964, **13**, 89.]
14. Heine, V., and Abarenkov, I., *Phil. Mag.*, 1964, **9**, 451.
15. Wilson, H. A., "The Theory of Metals." Cambridge University Press, London. 1953.
16. Hasted, J. B., "Physics of Atomic Collisions." Butterworths, London. 1964.
17. Rushbrooke, G. S., *Physica*, 1960, **26**, 259.
18. Wertheim, M. S., *J. Math. Phys.*, 1964, **5**, 643.
19. Barker, J. A., "Lattice Theories of the Liquid State." Pergamon, London. 1963. [A convenient discussion of the cell model and models of this type.]
20. Sjölander, A., *In* "Phonons and Phonon Scattering," (Ed. T. A. Bak), p. 76. Benjamin, New York. 1964.
21. Ashcroft, N. S., and Leckner, J., *Phys. Rev.*, 1966, **145**, 83.
22. Kahn, A., *Phys. Rev.*, 1964, **134**, A367.
23. Schofield, P., *Proc. phys. soc.*, 1966, **88**, 149.
24. Johnson, M. D., Hutchinson, P., and March, N. H., *Proc. R. Soc.*, 1964, **282A**, 283.
25. Gaskell, T., *Proc. phys. Soc.*, 1966, **89**, 231.
26. Lomer, W. M., and Low, G. C., *In* "Thermal Neutron Scattering," (Ed. P. A. Egelstaff), Chapter 1. Academic Press, London. 1965.
27. van Hove, L. *Phys. Rev.*, 1954, **95**, 249.
28. Blume, M., *In* "Symposium on Inelastic Scattering of Neutrons by Condensed Systems," p. 1. Brookhaven National Laboratory, BNL 940 (C–45), 1966.

29. Landau, L. D., and Lifshitz, E. M. "Electrodynamics of Continuous Media," Chapter XIV. Pergamon, Oxford. 1960. [A detailed discussion of the correlation function treatment of light scattering and visco-elastic theory is given by Rytov, S. M., *J. exp. theor. Phys. USSR*, 1957, **33**, 166, 514 and 671 (*Eng. Transl. Soviet Phys.*, 1958, **6**, 130, 401 and 513.)]
30. Placzek, G., *Phys. Rev.*, 1952, **86**, 377.
31. Schofield, P. *Phys. Rev. Letts.*, 1960, **4**, 239.
32. Uhlenbeck, G. E., and Beth, E., *Physica*, 1936, **3**, 729 and 1937, **4**, 915.
33. Egelstaff, P. A., *In* "Inelastic Scattering of Neutrons in Solids and Liquids," p. 25. International Atomic Energy Agency, Vienna. 1961.
34. Abragam, A., "The Principles of Nuclear Magnetism." Clarendon Press, Oxford. 1961.
35. Chudley, G. T., and Elliott, R. J., *Proc. phys. Soc.*, 1961, **77**, 353.
36. Vineyard, G. H., *Phys. Rev.*, 1958, **110**, 999.
37. Egelstaff, P. A., and Schofield, P., *Nucl. Sci. Engng*, 1962, **12**, 260.
38. de Gennes, P. G., *Physica*, 1959, **25**, 825.
39. Kubo, R., *In* "Statistical Mechanics of Equilibrium and Non Equilibrium," (Ed. J. Meixner), p. 81. North Holland, Amsterdam. 1965.
40. Sears, V. F., *Proc. phys. Soc.*, 1965, **86**, 953.
41. Nakahara, Y., and Takahashi, H., *Proc. phys. Soc.*, 1966, **89**, 747.
42. A summary of the neutron-scattering experiments of S. J. Cocking and P. D. Randolf on liquid sodium may be found in Egelstaff, P. A., *Rep. Prog. Phys.*, 1966, **29**, 333.
43. Rahman, A., *Phys. Rev.*, 1964, **136**, A405.
44. A convenient review of the theories of viscosity is given by Brush, S. G., *Chem. Rev.*, 1962, **62**, 513.
45. A convenient review of the theories of thermal conductivity is given by McLaughlin, E., *Chem. Rev.*, 1964, **64**, 389.
46. Glasstone, S., Laidler, K. J., and Eyring, H., "Theory of Rate Processes." McGraw-Hill, New York. 1941.
47. Cohen, M. M., and Turnbull, D., *J. chem. Phys.*, 1959, **31**, 1164.
48. Kirkwood, J., *J. chem. Phys.*, 1946, **14**, 180. ⎫ These three papers
49. Kirkwood, J., and Rice, S. A., *J. chem. Phys.*, 1959, **31**, 901. ⎬ are discussed in
50. Rice, S. A., *Molec. Phys.*, 1961, **4**, 305. ⎭ reference 5.
51. Frenkel, J., "Kinetic Theory of Liquids." Oxford University Press, London. 1946.
52. Lodge, A. S., "Elastic Liquids." Academic Press, London. 1964.
53. Green, H. S., "Molecular Theory of Fluids." North Holland, Amsterdam. 1952.
54. van Hove, L., *Physica*, 1958, **24**, 404.
55. Sjölander, A., *In* "Thermal Neutron Scattering," (Ed. P. A. Egelstaff), Chapter 7. Academic Press, London. 1965.
56. Egelstaff, P. A., *Rep. Prog. Phys.*, 1966, **29**, 333.
57. Chester, G., *Rep. Prog. Phys.*, 1963, **26**, 411.
58. Egelstaff, P. A., and Ring, J., *In* "Physics of Simple Liquids." (Ed. Temperley, Rowlinson and Rushbrooke), Chapter 7. North Holland, Amsterdam. 1967.
59. Hapgood, H. W., and Schneider, W. G., *Can. J. Chem.*, 1954, **32**, 98.
60. Bagastskii, M. I., Voronel, A. V., and Gutak, B. G., *Eng. Transl. Soviet Phys.*, 1963, **16**, 517.
61. Chase, C. E., Williamson, R. C., and Tisza, L. *Phys. Rev. Letts.*, 1964, **13**, 467.
62. Partington, J. R., "An Advanced Treatise of Physical Chemistry," Vol. II, "Properties of Liquids." Longmans Green, London. 1961.

63. Levesque, D., *Physica*, 1966, **32**, 1965.
64. Guggenheim, E. A., *J. chem. Phys.*, 1945, **13**, 253.
65. Fisher, M. E., *J. Math. Phys.*, 1964, **5**, 944.
66. Hemmer, P. C., Uhlenbeck, G. E., and Kac, M., *J. Math. Phys.*, 1964, **5**, 60.
67. Widom, B., *J. chem. Phys.*, 1965, **43**, 3898.
68. Pines, D., and Nozieres, P., "The Theory of Quantum Liquids," Vols. I and II. Benjamin, New York. 1966.
69. Jastrow, R., *Phys. Rev.*, 1955, **98**, 1479.
70. Wu, Fa. Y., and Feenberg, E., *Phys. Rev.*, 1961, **122**, 739.
71. Massey, W. E., *Phys. Rev.*, 1966, **151**, 153.
72. Henshaw, D. G., and Woods, A. D. B., *Phys. Rev.*, 1961, **121**, 1266.
73. Feynman, R. P., *In* "Progress in Low Temperature Physics," Vol. 1 (Ed. C. J. Gorter), p. 17. North-Holland, Amsterdam. 1955.

Subject Index

Spherical molecules, 2, 30
Spectral density, 135, 146, 184, 186
Strain tensor, 169
(rate of), 174
Stress
correlation function, 180, 182, 184, 187
tensor, 173ff, 179
Stationarity condition, 134, 139
Stochastic force, 127, 141, 143
Structure factor, 51, 73ff, 80, 111, 196, 210
Superfluid, 220ff
Superposition approximation, 56, 217
Symmetry of $G(\mathbf{r}, \tau)$, 108

T

Temperature fluctuations, 9, 171
Thermal
conductivity, 4, 5, 51, 159ff, 162, 178, 189
expansion, 178
Thermodynamic
fluctuations, 7, 114
limit, 12ff
properties, 78
Time
diffusive step, 129ff, 157
short in $G(\mathbf{r}, \tau)$, 113
substitutions, 109
Total correlation function, 56, 211
Transport properties, 10, 91, 148, 192
Triple point, 2, 19, 87, 145, 175, 199, 212
Two-fluid model, 223

U

$u(r)$
Buckingham, 28
definition of, 18
hard sphere, 28

Harrison, 47
Lennard–Jones, 28
real metal, 48
Ultrasonic data, 172

V

Velocity correlation function, 133ff, 155
Velocity of sound, 4, 162, 172, 202
Virial
coefficients, 23, 39
equation, 20
expansion, 22, 39
Visco-elastic theory, 166ff, 176, 184
Viscosity
shear, 4, 5, 148ff, 160, 166, 177, 183, 186, 221, 226
bulk, 160, 166, 177, 184, 186
Volume (or density) fluctuations, 9, 21, 93, 164, 200
Vortex lines, 222
Vibration, maximum frequency of, 107, 193, 195

W

Waves
longitudinal, 163, 164, 170ff
sound, 4, 156, 162, 164ff, 170ff, 195, 224
transverse, 163, 168ff

X

X-ray scattering, 66, 70, 97

Y

YBG (Yvon–Born–Green) equation, 55, 81, 84, 89, 204, 217
Yvon theorem, 140